JOURNAL OF
GREEN ENGINEERING

Volume 4, No. 4 (July 2014)

Special Issue on

Supporting Ecosystem Services through Green Engineering

Guest Editor:
Sjur Baardsen

JOURNAL OF GREEN ENGINEERING

Chairperson: Ramjee Prasad, CTIF, Aalborg University, Denmark
Editor-in-Chief: Dina Simunic, University of Zagreb, Croatia

Aims and Scopes
Journal of Green Engineering will publish original, high quality, peer-reviewed research papers and review articles dealing with environmentally safe engineering including their systems. Paper submission is solicited on:

- Theoretical and numerical modeling of environmentally safe electrical engineering devices and systems.
- Simulation of performance of innovative energy supply systems including renewable energy systems, as well as energy harvesting systems.
- Modeling and optimization of human environmentally conscientiousness environment (especially related to electromagnetics and acoustics).
- Modeling and optimization of applications of engineering sciences and technology to medicine and biology.
- Advances in modeling including optimization, product modeling, fault detection and diagnostics, inverse models.
- Advances in software and systems interoperability, validation and calibration techniques. Simulation tools for sustainable environment (especially electromagnetic, and acoustic).
- Experiences on teaching environmentally safe engineering (including applications of engineering sciences and technology to medicine and biology).

All these topics may be addressed from a global scale to a microscopic scale, and for different phases during the life cycle.

Published, sold and distributed by:
River Publishers
Niels Jernes Vej 10
9220 Aalborg Ø
Denmark

Tel.: +45369953197
www.riverpublishers.com

Journal of Green Engineering is published four times a year.
Publication programme, 2013–2014: Volume 4 (4 issues)

ISSN 1904-4720 (Print Version)
ISSN 2245-4586 (Online Version)
ISBN 978-87-93237-59-9 (this issue)

JOURNAL OF GREEN ENGINEERING

Volume 4 No. 4 July 2014

Foreword

Ecosystem Services have always been there as an environmental asset, serving humankind and nature itself. They are rooted in the stocks of natural resources, but as a concept, Ecosystem Services relate to the flow of services based on these stocks. The concept builds on several disciplines, where ecology is a major one, but also e.g. economics has been described as an important part of the concept over the last two decades. Subsequently, with economics follows technology and thus engineering by providing increasingly more effective technologies: Producing more output from a given set of inputs or producing the same output from less inputs. This represents a basic contribution to sustainable production. At the same time, engineering has also developed in a more broadly defined sustainable direction through the concept of Green Engineering. This concept is thus not only about economic efficiency, but also relates to a broad range of environmental and social issues, as an answer to the modern concept of sustainability. Nevertheless, Green Engineering has so far not been explicitly integrated into the concept of Ecosystem Services. This Special Issue of JGE is an attempt to do so.

The concept of Ecosystem Services has become increasingly widespread and commonly used in scientific literature since it was established 35 years ago. Nevertheless, the current applied definition of the concept is relatively new. It was formalized by the United Nations 2005 Millennium Ecosystem Assessment (MEA). UN grouped Ecosystem Services into four broad categories: *Supporting services*, such as seed dispersal, crop pollination and primary production; *Provisioning services*, such as the production of food, clean water, hydropower and timber; *Regulating services*, such as the control of carbon sequestration and climate, waste decomposition and detoxification, purification of water and air, water quality and diseases, pest and disease control; and *Cultural services*, such as spiritual and recreational benefits. The MEA was a broad synthesis involving over 1300 scientist and providing guidelines for decision-makers.

The concept *Ecosystem Services* thus embraces products obtained from ecosystems, benefits obtained from regulating the ecosystems, non-material benefits people obtain from ecosystems through e.g. recreation, and aesthetic experiences, and the concept also includes services that are necessary for the production of all other Ecosystem Services.

In this special issue, we invited papers that deal with how Green Engineering may contribute to and enhance Ecosystem Services. We did so being well aware of the challenges inherent in the fact that this is a new approach, trying to bring together the two established concepts Green Engineering and Ecosystem Services. The result is promising, although we note that this is new to the scientific society, and thus that submitted papers tend to integrate the concepts less than we may had hoped for. Nonetheless, we regard this an important first step in the direction of more integrated research with wider scopes. We also note that the majority of the papers come from Forest Engineering. This is natural, as the forest ecosystems dominate in Europe, and this Special Issue deals with the engineering aspects. Finally, we also received papers on policy development and implementation of policies related to Ecosystem Services.

The Special Issue sets out with a paper by Adrian Enache and Karl Stampfer (Machine utilization rates, energy requirements and greenhouse gas emissions of forest roads construction and maintenance in Romanian mountain forests). They assessed the environmental footprints of forest roads in terms of embodied energy and greenhouse gas emissions due to construction and maintenance, applying Life Cycle Assessment and an input-output model for two study areas. They measured energy input and calculated CO_2 equivalents, and found that although road construction and maintenance are important sources of Green House Gas (GHG) emissions, these emissions are very minor when compared to the CO_2 equivalents stored in the growing stock of the opened forest areas. Terrain characteristics showed a strong influence on the amount of fuel consumption, on required energy input and on GHG emissions, leading to higher environmental burden and higher road construction costs.

Jonas Lindström and Dag Fjeld follow up with a somehow related paper on transport in forests (A process perspective on the timber transport vehicle routing problem). They map how the timber transport vehicle routing problem is solved in practice, and which consequences different ways of solving the problem has for service and efficiency. Fifteen haulage contractors were selected for the mapping. The results show, i.a., how net operating margins decrease with increasing levels of supplier service, and that those using a simplified process model have much higher net operating margin than those using a more complete process model.

The third and last paper related to forestry is written by Bruce Talbot and Rasmus Astrup (Forest operations and ecosystem services in Norway – a review of the issues at hand and the opportunities offered through new technologies). Their point of departure is the fact that forest operations are the most costly part of forest management and, at the same time, can be traced to most of the negative externalities on the environment, often with strong visual impacts, especially in steep terrain. Their paper reviews emerging technology-based engineering solutions that may reduce the impact of forest operations on the environment while increasing the efficiency of operations, resulting in an overall higher level of forest Ecosystem Service provision. They point at advances in airborne laser scanning and forest machine control and automation systems, and the availability of remotely-sensed high-resolution data, which now provide considerable potential to improve the management and precision of forest operations, and thus mitigating environmental damage.

Neil McIntosh and Tamsin Burbidge (The Natura 2000 biogeographical process) contribute with a communication on the Natura 2000 biogeographical process. They give an overview of the background, purpose and core messages of the process, details the strategic content, aims and opportunities and describes both the current participation and the future plans.

Kristijan Èiviæ and Laurence M. Jones-Walters (Implementing green infrastructure and ecological networks in Europe: lessons learned and future perspectives) conclude this Special Issue. They emphasize the recent development from regarding Ecological Networks as pure migration paths in fragmented landscapes to Green Infrastructure by adding social and economic aspects. This also opens for a much wider inclusion of "Green Engineering" aspects and relates well to Ecosystem Services. The authors emphasize that further work should consider the practicalities of the full translation into integrated functional ecological networks and making them building blocks of the green infrastructure both at the level of policy and practice. Moreover, they stress the importance of information and communication.

Sjur Baardsen
Norwegian University of Life Sciences,
Norway.

The Natura 2000 Biogeographical Process

Neil McIntosh and Tamsin Burbidge

ECNC-European Centre for Nature Conservation and ECNC Land & Sea Group
PO Box 90154, 5000 LG Tilburg, The Netherlands
Corresponding author: Neil McIntosh <McIntosh@ecnc.org>

Received 2 June 2014; Accepted 27 January 2015;
Publication 19 March 2015

Preface

The Natura 2000 Biogeographical Process was launched by the European Commission in 2011 to assist Member States in managing Natura 2000 as a coherent ecological network. This article provides the background, purpose and core messages of the Process, details the strategic content, aims and opportunities and describes both the current participation and the future plans.

Background and Purpose

The Natura 2000 Biogeographical Process was launched by the European Commission in 2011 to assist Member States in managing Natura 2000 as a coherent ecological network. The primary purpose of the Natura 2000 Biogeographical Process is to assist Member States to meet legal obligations under the nature directives with respect to the favourable conservation status of habitats and species of community interest. Through the Natura 2000 Biogeographical Process, a key aim is to ensure that Member States and expert stakeholders are enabled to realise collaborative networking events, associated information sharing and cooperative knowledge building activities, linked to common strategic priorities. The EU 2020 Biodiversity Strategy calls for a step change in efforts to halt the loss of biodiversity and to restore essential services that a healthy natural environment provides.

Journal of Green Engineering, Vol. 4, 261–270.
doi: 10.13052/jge1904-4720.441

Core Messages of the Process

The following points highlight key features of the Natura 2000 Biogeographical Process:

- Participation in the Natura 2000 Biogeographical Process is voluntary;
- The Process provides a valuable means to work collectively towards achieving the legal obligations of the Nature Directives;
- The Process offers a practical framework for networking, sharing information and experience and building knowledge about the most effective ways to reach and maintain favourable status for habitats and species of European Community importance - this includes opportunities to identify and promote the multiple benefits linked to such actions;
- The Process focuses on practical habitat management and restoration activities and provides a framework to share best practices, compare approaches, build contacts, exchange information and build new knowledge;
- The Process is supported by follow-up networking events designed to further build practical knowledge and capacity, along with a dedicated Natura 2000 Platform to communicate and share information.

Strategic Content

As responsibility for implementation of Natura 2000 and ensuring progress towards the EU's Biodiversity Strategy targets lies with Member States, they are key actors in the Natura 2000 Biogeographical Process. Also, there are significant opportunities through the Process to improve mobilisation of expert networks and inputs from other key stakeholders. This is important to benefit from the direct experience of Natura 2000 practitioners, expert stakeholders and Member States' representatives with specific responsibilities for implementation of Natura 2000. This underlines the strategic and operational importance of the Process and the integrated inputs required from diverse actors.

In the Natura 2000 network, as in the Strategy, the needs of biodiversity are central, but not isolated – for example, by taking an ecosystem-based approach (i.e. recognizing the full array of interaction within an ecosystem rather than focusing on one specific issue, species or ecosystem service), it is possible to ensure that nature (and therefore Natura 2000) continues to contribute to growth at local, regional, national and European levels. Working through the cooperation mechanisms provided by the Process, this means that:

- The inputs of Member States and expert stakeholders are central to defining the forms of collaboration required to achieve the 2020 targets.
- Collaborative actions should focus on nature's many processes and functions to improve habitat condition and generate multiple benefits, including social prosperity and welfare.
- There is opportunity to reflect and think collectively about practical ways to improve the favourable conservation status of habitats and species and to learn from Article 17 reporting experience – this includes utilizing the Process to, for example:
  - Identify the forms of collaboration appropriate for agreed common priorities, including exploring scope for potential LIFE or Interreg project proposals;
  - Discuss, agree and set conservation objectives at biogeographical level;
  - Define favourable reference values for conservation status at different levels within a biogeographical region.

Therefore, the Strategy captures the common objectives and specifies the key targets to be met - for example, to build understanding about how EU 2020 BD Strategy Target 1 is interpreted. Working with expert stakeholders in the NGO community, over the last 20+ years, significant gains in ecology knowledge, information and practical management experience have been acquired about Natura 2000. However, especially with regard to strengthening Natura 2000 as a coherent ecological network, there is scope to generate measurable improvements about how this knowledge and experience can be collectively developed and collaboratively applied. Cumulatively, improved nature conservation management practices will enable greater progress to be achieved towards the 2020 Biodiversity Strategy goals and targets. In this way, the Process can increasingly be utilized to guide participants towards "common directions".

Aims and Opportunities

To achieve progress within the strategic context, it is essential that there is clear and shared understanding about what is already known, what has to be achieved and what actions require to be developed together to safeguard biodiversity in Europe. The Process provides practical means to exchange the information, experience and knowledge that is required to identify and define common solutions and develop cooperative actions, which can be delivered to ensure progress towards the EU 2020 Biodiversity targets. Specifically, within the Natura 2000 Biogeographical Process, there are opportunities to generate better integrated ecosystem-based management approaches, through mobilizing greater inputs from strategic stakeholders and increasing participation from practitioners. There are support opportunities to network (for example, through workshops or working groups) that generate recommendations for practical Natura 2000 management, matched by shared commitments for future actions.

Working with established Steering Committees and experts in each biogeographical region is designed to foster greater focus on strategic targets – for example, explore how to use Article 17 data more proactively and build common understanding about core strategic policy areas such as interpretations of favourable conservation status and favourable reference values. Also, through the Process, there is an important opportunity to catalyse change, by increasing practical know-how and insights on the basis of learning from pilot studies: this includes, benefiting from experience gains realized through LIFE (& other) projects and monitoring results (including Article 17 reporting) to increase synergies; at the same time, for example, through improved information sharing about management practices and approaches per habitat type, and promoting identified best practices for specific Natura 2000 management issues. The aim is to continue to develop the Natura 2000 Biogeographical Process with greater focus on strategic outcomes achieved through supporting a range of practical management cooperative actions that strengthen the implementation of Natura 2000.

Current Participation

One of the key features of the Natura 2000 Biogeographical Process is that participation is voluntary. Involvement to-date has been very positive. The data presented in the tables and in Figure 1 has been compiled from all biogeographical regions since the start of the process in 2011 – these include all policy makers, experts, consultants, site managers, NGO representatives and other stakeholders involved in either the coordination of the process, the Steering Committee, expert contributions to the Background Document, preparatory workshops and seminars.

The information used to compile this overview of the Natura 2000 biogeographical process has been extracted from the Natura 2000 Platform database, developed and maintained by ECNC. As can be seen, the main interest in the Process is predominantly around the location of the central administration. The high number of participants from North, West and Central European countries

Table 1 Number of active participants in the process by biogeographical region

Biogeographical Region	Active Participants
Alpine	329
Atlantic	255
Boreal	194
Continental	110
Macaronesian	2
Mediterranean	161
Marine	2

This table shows the number of active participants involved in the process by Biogeographical region. These figures include members of the Steering Committees, experts contributing to background documents, and participants in the various events (workshops, seminars, ad hoc meetings, networking meetings, etc). A totals row is not included in this table, because many individuals participate in more than one Biogeographical process. This information is also illustrated in the following pie chart:

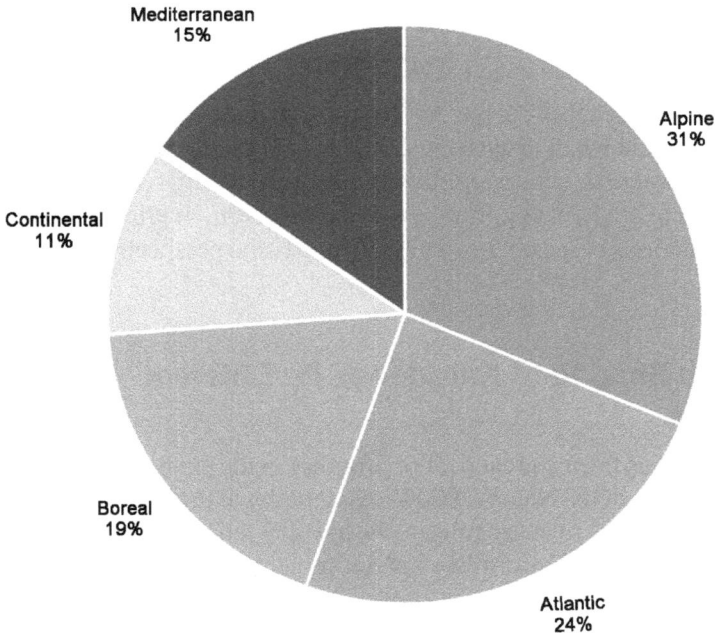

Figure 1 Number of active participants in the Natura 2000 biogeographical process since 2011.

reflects the fact that the first three Biogeographical Regions involved in the Process are Boreal, Atlantic and Alpine. The Mediterranean Biogeographical Process started in May 2014, and there has been significant interest and levels of participation. The sectors from which participants have been engaged are as follows:

Sector	# People
Agriculture	35
Environment	398
Financial	5
Fisheries	4
Forest	46
Hunting	7
Industry	3
Nature conservation	784
Planning	108
Site management	64
Tourism	5
Transport and infrastructure	8
Water	32
Total	1499

There is growing cross-sector involvement in the Process, which is fundamentally required when discussing challenges and issues related to Natura 2000 and, in particular, ecosystem-based management approaches. Through the Process, there are new opportunities to define common priorities and take collaborative actions based on the integrated needs and perspectives of diverse stakeholders.

Progress in Sharing of Knowledge for Different Ecosystems

In general there has been a great deal of progress in the sharing of knowledge since the beginning of the Natura 2000 Biogeographical Process. Each biogeographical region has its own specific needs, but also scope to address common themes, such as access to information about project funding opportunities.

Within the Boreal region much progress in development of discussions and sharing of knowledge has already been made. Many seminars and workshops have been organised including a wetlands restoration workshop, which allowed for the sharing of strategies and experiences between Member States and expert stakeholders. Also, the exchange of best practices for forestry between Member States and stakeholders is currently receiving attention. Further plans for sharing of knowledge include a seminar to address common issues within grassland habitats and workshops about good practices.

Within the Atlantic region progress includes developments in the problem of Nitrogen deposition and diffuse pollution of water. A knowledge exchange meeting was organised to determine: CL's/per habitat and nitrogen ceilings needed to reach FCS. Translation of the OBN/government report on nitrogen deposition and diffuse pollution of water (including problems and measures) into English has also been carried out. Evaluating utility of AERIUS model from the Netherlands in other communities is ongoing and will result in the circulation of information about the model to all Member States. Further actions to be taken include promoting the development of a mobile app to assist the public to be aware of Natura 2000 sites and provide information including for instance (common) species and Nitrogen deposition. This will include a meeting of Member States to identify existing IT applications and how these can be more widely developed and applied.

Much progress has been made with the sharing of knowledge in the Alpine region including the establishment of a sustainable forum and workshops with field trips (every 2 years) to share knowledge on river and lake restoration and maintenance: this series of networking and knowledge-building events will also provide scope to discuss common issues on the impacts of, for example, hydropower and conservation status. This links to other current activities, including an imminent EC guidance document on hydropower and Natura 2000: a draft guidance paper on flood prevention and maintenance works for river ecosystems is currently in preparation and the work of the Alpine Process participants for Natura 2000 habitats and species in their Biogeographical region is contributing to that. Further actions for other ecosystems include raising awareness of grassland habitat conservation issues by organizing events and sharing information on both grassland conservation and ecosystem services.

Future Plans

Following the success of the first seminars (Boreal Biogeographic region in May 2012, Alpine Biogeographical region in November 2013, Atlantic Biogeographical region in December 2013 and Mediterranean Region in May 2014), the Natura 2000 Biogeographical Process continues, with further thematic 'Kick-off seminars' and 'Review seminars' being planned. Each seminar will be informed by background information on the conservation status and needs of the selected habitat types and species: a list of related habitat groups, crosscutting issues and problems whose solutions should directly contribute to achieving favourable conservation status will also be addressed.

In addition, the seminars will be organised at the level of biogeographical regions at intervals which take stock of the results of the thematic events in the region. Follow-up actions, identified as being most useful to Process stakeholders, can be further developed through networking and cooperation under a new proposed Natura 2000 Biogeographical Networking Programme. The follow-up actions can take the form of conferences, workshops, expert meetings, or study visits.

Hosted by national or regional actors (rather than lead countries), the Seminars will be supported, organized and facilitated by the EC's contractor, under a technical assistance contract to provide added value opportunities that progress the favourable conservation status of habitats and species of Community interest. The Seminars and follow-up networking events will aim to result in a jointly agreed list of recommendations and priority actions identified by Member States and expert networks for follow-up in-depth cooperation, networking and collaborative action in respective regions and, where appropriate, also between regions.

The results of the Seminars and networking events will continue to be shared on the **Natura 2000 Platform**. As the content of the Platform expands with greater volumes of relevant Natura 2000 information, it will continue to be developed as a web-based tool for networking, dialogue building and exchanging information on conservation objectives and measures between all actors involved in the Process.

Biographies

Neil McIntosh studied English Language & Literature, Politics and Linguistics, graduating with an MA (awarded with merit) from the University of Edinburgh in 1988. He is a qualified project management practitioner, trainer and assessor of training courses. Neil has worked in the public, private and voluntary sectors in the fields of environment, natural heritage, health and business development. Living in the Netherlands for almost 10 years, he is fluent in Dutch. Currently, Neil is the Deputy Executive Director and Head of Strategy and Innovation for ECNC- European Centre for Nature Conservation.

Tamsin Lucy Burbidge graduated from the University of Manchester in June 2009 with a BSc. in Zoology with Industrial Experience (Hons.). She successfully completed an MSc. in Biology at the University of Leiden, the Netherlands. Tamsin continues to live in the Netherlands and has been working as volunteer at ECNC- European Centre for Natura Conservation, based in Tilburg, for almost one year. She has worked as a research assistant in Europe and Africa on various aspects of zoology and has completed research internships on conservation, behavioural and population biology.

Forest Operations and Ecosystems Services in Norway – A Review of the Issues at Hand and the Opportunities Offered through New Technologies

Bruce Talbot[1] and Rasmus Astrup[2]

[1]Department of Forest Technology, Norwegian Forest and Landscape Institute, P.O. Box 115, 1431 Ås, Norway
[2]Department of Forest Resources, Norwegian Forest and Landscape Institute, P.O. Box 115, 1431 Ås, Norway
Corresponding Authors: {bta;raa}@skogoglandskap.no

Received 17 October 2014; Accepted 27 November 2014;
Publication 19 March 2015

Abstract

It is widely recognized that forests should be managed *inter alia* for the provision of timber, biomass for energy, bio-chemicals, biological diversity, carbon storage, water purification, outdoor recreation and other ecosystem services. Forest operations are the most costly part of forest management and at the same time can be traced to most of the negative externalities on the environment often with strong visual impacts, especially in steep terrain. This paper reviews emerging technology-based engineering solutions that may reduce the impact of forest operations on the environment while increasing the efficiency of operations resulting in an overall higher level of forest ecosystem service provision. Advances in forest machine control and automation systems, and the availability of remotely-sensed high resolution data now provide considerable potential to improve the management and precision of forest operations. Improved planning procedures and more precise operations offer a considerable opportunity for mitigating environmental damage. Accurate positioning of operations machinery allows for the generation

Journal of Green Engineering, Vol. 4, 271–290.
doi: 10.13052/jge1904-4720.442

of automated warnings in the case of transgressing property or key habitat boundaries. Terrain models derived from airborne laser scanning (ALS) data have been shown to be useful in locating effective extraction trails or timber landings, thereby increasing efficiency and reducing site impact such as rutting and compaction.

Keywords: Timber harvesting, logging, forest operations, environmental impact.

1 Introduction

Norway has an area of 8.2 million ha of productive forest [1] and a low population density (14 people per km^2), and has similar expectations for the provision of ecosystems services from forests as do highly populated countries like Germany (230 people per km^2) and France (121 people per km^2) [2]. A recent report commissioned by the Norwegian Ministry of the Environment emphasises the importance of forests managed *inter alia* for the provision of timber, biomass for energy, bio-chemicals, biological diversity, carbon storage, water purification and outdoor recreation [3]. These expectations with regard to forest based services are not unique to Norway, but are in fact universal [4].

Primary energy supply and land use scenarios envisage a future that is more heavily reliant on terrestrial ecosystems to supply food, fiber, and energy [5, 6]. The scale of the transformation challenge is enormous in a world where land resources are already scarce [7]. It is widely recognized that an increased reliance on forest for the provision of fiber will pose a threat to biodiversity and other ecosystem services [8]. Besides, an active and profitable Norwegian forest sector with a competitive forest industry is essential for rural development in large parts of the country [9]. The government intends to strengthen the value-adding capability of the forestry sector, with the goal of mobilizing resources for commercial development and climate change mitigation [9]. The current annual volume of roundwood sales in Norway is about 9 mill m^3 but an annual growth of almost three times that indicates a significant potential for increased procurement of Norwegian forest fiber [1].

Norway has a high-cost society and the forest sector is struggling to stay competitive in the global market. Despite phenomenal efficiency gains over the past half-century, logging remains the most cost intensive part of forest management [10]. Forest operations are not only costly, they also cause the largest negative externalities on the environment and have the strongest visual impact [11], especially in steep terrain. The public debate related to the environmental impact of forestry is intense. As observed in other parts of the world, this poses a threat to the social license to practice [12] leading to political and public resistance towards further development of the forest-based bio-economy. Hence, when considering the development of the Norwegian forest sector as an economic, social and environmentally sustainable pillar of the bio-economy, a strong emphasis should be placed on the economic and environmental efficiency of harvesting operations. Environmental efficiency comes about through minimizing environmental damage for a given production output. However, environmental efficiency studies are virtually non-existent in forestry, but are more common in agricultural production settings [13–15].

Forest engineering, applied as a professional scientific discipline, involves both developing and maintaining the infrastructure necessary in accessing forests for any purpose, as well as designing tools, machines, machine systems and work methods for carrying out forest management tasks. Forest engineering, as a sub-discipline of environmental engineering, and the way in which it is practiced, is therefore central to any debate regarding the goods and services produced on forest land. The two leading forest certification schemes in Norway; the Forest Stewardship Council (FSC) [16] and the Programme for the Endorsement of Forest Certification (PEFC) [17] have developed a set of sustainability indicators that include elements related to forest operations. The Norwegian association of heavy equipment contractors (MEF), which includes timber harvesting contractors, are full members of PEFC, and their members are therefore obliged to follow the guideline set by the PEFC 'Living Forest Standard' which was drafted by a range of organisations representing outdoor recreation, forest owners, the Norwegian biodiversity network, industrial associations, and WWF Norway [18]. The standard states that the harvesting potential should be exploited within a framework set by economic, biodiversity and other environmental considerations and values. In addition the standard maintains that a higher share of selection cutting should be carried out through thinning, and the selection of harvesting methods and the implementation of the harvest should be adapted to local conditions, maintain local environmental qualities, take due consideration of the landscape and ensure conditions for satisfactory regeneration with species suited to the local growing area [18].

Issues of direct relevance to the way forest operations are carried are:

- Requirements on the workforce and their qualifications: Training courses run continuously at the Norwegian Forestry Extension Institute and skill development in steep terrain logging documented through research [19, 20].

- Waste management, here especially engine and hydraulic oil, spares, steel cable, chains etc., as well as a requirement to use the best available technology (BAT).
- Key habitats: The management of at least 5% of productive forest as areas of ecological importance, requiring the documentation and registration of these areas. With regard to mountain forests, the standard requires that the harvesting methods for spruce must follow the "mountain forest selection cutting system" as far as possible. Small-scale clear-cutting and smaller seed tree stand felling should be used as far as possible to promote regeneration in the pine forest. The importance of protection forests against landslides in this area has recently been addressed [21].
- Outdoor recreation: Forest management activities must consider maintaining the quality of outdoor experiences, especially along hiking and skiing trails. All commercial activity in forest areas must be conducted in such a way that the *de facto* content of the right of unimpeded access is maintained.

Wheel tracks that cause water runoff and erosion, transport damage to hiking and ski trails and other significant damage must be repaired as soon as moisture conditions permit once the extraction trail is no longer in use. Excessive wheel rutting (Figure 2, right) and the placement of harvesting residues are two of the most common problems experienced by forest visitors.

While the Norwegian forest sector is appreciably smaller than those of neighbouring Sweden and Finland (which harvest around 80 and 70 million m^3 a^{-1} respectively), much of the forest area is found in steep, difficult, or highly susceptible terrain, providing a challenge to forest engineering in suggesting environmentally acceptable solutions. This paper builds on the hypothesis that environmental damage directly or indirectly impairs the capacity of the forest to provide the ecosystems services desired by society. Only the most obvious physical forms of environmental or ecological damage are included in this paper, and these are categorised below as a frame of reference:

- Efficiency of resource utilisation and damage to key woodland habitats
- Soil and Water: rutting, compaction, surface and gulley erosion, turbidity and sedimentation
- Aesthetic and cultural heritage impacts, as well as impacts on the recreational use of forests

The paper focuses on a review of emerging technology-based solutions that may reduce the impact of forest operations on the environment while increasing the efficiency of operations. Hence, the paper does not address issues around changes to ecosystem functionality, neither does it consider water table fluctuations or influences on bio-geochemical cycles (including carbon fluxes) from the utilisation of various tree parts or harvesting regimes.

2 Key Organisational and Technical Advances Available to the Forest Operations Manager

Advances in forest machine control and automation systems, and the availability of remotely-sensed high resolution data now provide considerable potential to improve the management and precision of forest operations [22]. Improved planning procedures, and more precise operations offer a considerable opportunity for mitigating environmental damage. The utilization, productivity and economic efficiency of forest machines, and meeting of specific product orders, is fully dependent on planning both within and between stands [23]. The direct cost of relocation alone can account for 10% of the harvesting price [24], which could be reduced through improved scheduling [23, 25]. In scheduling, machines can be ranked and allocated to stands on a broad range of environmental and performance criteria, including the interacting relationships of harvest season, soil texture, bulk density, surface soil strength and position in the landscape [26]. Furthermore, planning a harvest sequence at a landscape level with the inclusion of adjacency constraints, can reduce visual impact, erosion, landslide or avalanche danger, and enhance biodiversity [27]. However, advances in planning are wholly reliant on technical developments that are able to capture or convert large amounts of data into improved operational information. The following subsections will attempt to demonstrate the importance of the data collected and highlight applications of this data in emerging research.

2.1 Efficient Resource Utilisation and Avoidance of Damage to Key Woodland Habitats

Forest machine performance can be captured continuously using a combination of on-board productivity monitoring systems, including the StanForD data management system [28], CAN-bus machine performance data, and external sensor data [29, 30].

Accurate positioning of base machines allows for the generation of automated warnings in the case of transgressing property or key habitat boundaries, and also allows for machines to accurately log information on residual trees or other important sites or cultural heritage vestiges identified during harvesting. Accurate positioning of forest machines on steep slopes of fjords at high northern latitudes remains a significant challenge in carrying out precision forestry operations as both the azimuth and limited time windows for satellite connection present difficulties. Global Navigation Satellite System (GNSS) information can be enhanced with terrain surface information [31], and possibly supplemented with other positioning techniques such as laser [32] or visual telemetry based solutions [33], simultaneous location and mapping (SLAM) [34], and inertial navigation systems (INS) [35].

Terrain and tree size and species distributions have the largest effect on productivity [36] and can now be remotely assessed through airborne laser scanning (ALS). Stand-level bucking, which allows for detailed product recovery to be calculated from ALS data provides for a more accurate selection of stands for harvesting, both in time and space [37], and could therefore be an important parameter in future studies on environmental efficiency, where a detailed product output portfolio could be held up against an index of anticipated environmental damage [14]. The inclusion of stand level bucking prognoses from terrestrial scanning [38] or from inventory data, in completely new approaches to tactical planning [39, 40] potentially represents an important shift toward value maximization planning. As bucking patterns both strongly affect production [41] and transport costs and energy consumption [42], they play an important role in the entire procurement system [43]. The development of more agile, value maximizing supply chains could play an important role in revitalizing an economically and environmentally efficient Norwegian timber harvesting sector.

2.2 Soil and Water

Forests in wetter climates provide a stabilizing buffer in the hydrological cycle, reducing surface run-off and its associated erosional energy [44], regulating release to waterways [45] and promoting percolation of uncontaminated water to aquifers, as forest soils are generally subjected to fewer or no chemicals or fertilizers [46]. Impacts on soil and water related to forest engineering largely arise from both the construction and maintenance of forest infrastructure and from machinery in the harvesting and extraction of timber. With regard to forest roads, the Norwegian Ministry of Environment (MoE) report on Ecosystems

Services [3] highlights one of many ecosystem service trade-off dilemmas arising from joint ambitions of utilisation and protection, as the encroachment of roads is criticised as being detrimental to biodiversity and the wilderness experience through forest segmentation on the one hand, while their absolute necessity in facilitating a rational utilisation of forest products, as well as the access they provide to the recreational use of forests and mountain rangelands above the forest line is highlighted on the other.

The Norwegian forest road network remains sparse in relation to comparable countries, and is generally below 10 m ha^{-1}, with only 3–5 m ha^{-1} in the steep terrain in the coastal area. In total, some 400 000 ha of mature timber are located more than 2 km from a forest road, 728 000 ha between 1 and 2 km, and 782 000 ha between 500 m and 1 km from a road [1]. As road transport is at least 20 times more efficient than terrain transport [47] there is a strong incentive to expand the forest road network. The use of modern technology such ALS data help with detailed planning of forest roads, ensuring that they are as environmentally benign as possible. Examples of this are given by Aruga et al [48], who minimize sedimentation through improved road design, and Contreras et al [49] who apply road design geometry to an ALS derived DEM in estimating earthwork requirements. Related work employing ALS includes the mapping of gullies and headwater streams [50] and the estimation of gully erosion [51]. In Norway, increased rainfall intensity and magnitude has resulted in a recent revision of the forest road construction standards, stipulating amongst other things, changes to the frequency and dimension of drains [52].

As forests on the Norwegian west coast constitute an area of high rainfall (>2000 mm a^{-1}) with steep terrain (15 million m^3 of mature timber on slopes over 50% [1] and a low forest road network, they simultaneously offer a wide range of ecosystems services and are themselves highly susceptible to inadvertent damage through harvesting practices. Access to these stands is gained either by the construction of temporary skid trails which allow access to ground based fully mechanized systems (harvester-forwarder) (Figure 1, left) or by cable-yarders specially constructed on terrain-going base machines enabling them to work at some distance from a forest road (Figure 1, right). While cable-yarders minimise impact to the slope by partially suspending the trees during extraction, skid trail construction is a matter of some environmental concern [53] and a novel method of monitoring the extent of soil displacement arising from it has recently been tested in this area using drones [54]. Skid trails can be de-activated through a number of means with varying results [55, 56] and at various costs [57], although a reduced need for a skid trails in conjunction with improved planning procedures is the preferable option [58].

Figure 1 (left) construction of temporary skid trails in steep terrain and (right) specially developed Owren 350 terrain-going tower yarder for accessing sites with low bearing capacity or some distance from the forest road network.

Norway is renowned for its salmonid populations, which are associated with almost all the country's watersheds. Some 330 tonnes of wild salmon, sea trout and migratory char were caught by recreational anglers in 2013 [59], making fishing a significant contributor to the nature-based tourism portfolio. Sediment from forest roads and forest operations is known to impact on salmon populations through the silting of spawning beds [60].

Terrain models derived from ALS data have been shown to be useful in locating effective extraction trails [58] or timber landings [61], thereby increasing efficiency and reducing both site impact like rutting and compaction [62] as well as emissions [63] arising from unnecessary terrain transport. Detailed surface features not distinguishable from ALS data with a resolution of 2–5 points m^{-2} need to be quantified with specialized sensors, e.g. monitoring pitch and roll [64, 65], or can also be assessed post-harvest using drone based photogrammetry [54]. Tracks fitted to the bogie wheels increase machine flotation and reduce soil disturbance [62]. Apart from correct tire selection and inflation [66], recent engineering attempts at further reducing the impact include an extension of the length of the bogie axle itself [67] and adding another wheel set to the machine [68].

While technical developments provide essential tools for mitigating environmental damage, operator effect has been shown to explain up to 37% of productivity variance between machines [69] and is likely to account for an even larger share of site damage, although this has not been quantified. Estimates of the overall magnitude of the operator effect have improved through more specific analyses [70, 71], but the absence of explanatory

Figure 2 Forwarder fording a stream in a mountainous area (left), and (right) longer term consequences of heavy vehicle traffic during the spring thaw – here 1 year after harvesting.

factors detailing this effect represents a knowledge gap which will need to be addressed in minimizing the environmental impact of heavy forest machinery.

2.3 Aesthetic and Cultural Heritage Impacts, as well as Impacts on the Recreational Use of Forests

There has been some activity in remotely detecting and mapping cultural and environmental parameters. For cultural parameters, the use of ALS to detect archeological features in forests has led to considerable increase in the detection of important features [72]. Harvesters and Forwarders are equipped with monitors running mapping applications, where most are issued with layers representing the harvesting site, location of retention trees, or other areas that must be avoided. Cultural heritage sites are generally included in the information provided to the machine operator. However, the effectiveness of the approach is dependent on the accuracy of the GPS signal, and the detection of archeological features needs to be complemented with machine-mounted sensors to avoid damage to cultural heritage sites.

While there is broad agreement that the spatial configurations and properties of the landscape interact with each other [73], research work on internalizing the effect and monetary value of aesthetics has been limited in recent decades. A review of over 30 years of research considering the public preferences for forest structures, showed that people valued stands with a mixture of trees of different sizes and that larger clear cuts and obvious traces from forest operations were not appreciated [74]. However, while steep terrain harvesting is highly visible from a distance (Figure 3) the same study did point out that recreational tourists in the forest did

Figure 3   (left) A rare example of positive aesthetics from forest operations -the torch bearer mascot of the 1994 Winter Olympics at Lillehammer, created using cable-yarder harvesting on a steep mountainside, and (right) highly visible harvest site showing ravine landslide to the right.

appreciate openings that provided scenic views. Other studies describe the importance of both forest and individual trees that provide environmental services from a symbolic, cultural or spiritual perspective [75]. The fact that the retention of selected trees is also an important conservation measure in managed forests [76] re-emphasizes the need to apply precision forestry techniques, especially with regard to machine positioning and single tree management.

Outdoor recreation is an important part of the tourism industry in rural Norway, and can primarily be categorised as: freshwater fishing, hunting, backcountry hiking and skiing, and adventure activities [77]. Forest operations, especially wheel rutting, piling of harvesting slash and damage, or the closure of access roads can impede access to the area or detract from the backcountry experience, and are specifically highlighted in the 'Living Forests' standard [18]. Knock-on effects, such as a downturn in salmonid populations due to sedimentation of headwaters, can also have serious consequences for the nature-based tourism industry [60]. But while all Nordic countries do have specifically focused strategies or programs for outdoor recreation and nature-based tourism, there are currently no good indicators for a sustainable recreational use of forests or other natural resources [78]. However, some of these indicators could be recorded by the machines themselves. Data systems

(CAN-bus) on modern mechanised harvesting systems provide the capability to estimate wheel rutting through rolling resistance [29] and when augmented by sensors such as laser rut measurement systems (LRMS) and enhanced machine positioning could ultimately provide detailed maps of soil impact for amelioration or monitoring. Finally, improved data and techniques for multi-criteria harvesting planning, which takes into account edaphic, biological, hydrological and stakeholder information offers forest operations managers with a holistic management tool for largely avoiding both environmental and societal conflict while meeting economic goals [79].

3 Concluding Remarks

The availability of high resolution environmental data have become commonplace, as have a plethora of low-cost, rugged sensors that can provide visual, metric, chemical and positional data pertaining to the machines and their specific operations. Access to high precision geological and topographic maps provide operation managers at all levels with detailed information which can be used in avoiding damage to the forest. The use of drones and close range photogrammetry from consumer grade cameras allow for three-dimensional data to be captured and used for monitoring, for compliance with certification standards. Improved positioning systems can both provide the machine operator with information on what to cut and what to avoid, but also to flag waypoints indicating special areas of interest discovered during harvesting. New regulations on road construction and maintenance take regard for a changing climate with more intense precipitation and ensure that water is not channelled in a way that leads to landslides or other large scale damage. Forest machines fitted with high flotation tracks provide improved access but still exceed the bearing capacity of soils in an increasingly milder winter climate. The potential for monitoring environmental performance with machine-mounted sensors is at a very early stage, but it is expected to improve rapidly as low cost sensors become commonly available, allowing for remote 3rd party assessment of compliance with environmental standards. Ultimately, it is the machine operator who has the final capability to assess the situation in forest stands, and the responsibility to make the correct decision both with regard to the working method and the eventual suspension of operations in situations where damage to the forest obviously exceeds what could be called the social norm among professional forest engineers.

4 Acknowledgements

The authors wish to acknowledge the financial support received from the ECOSERVICE and *Sustainable utilization of forest resources in Norway* projects, funded by the Norwegian Research Council (NFR 233641/E50 and NFR 225329/E40) as well as the support from the Norwegian Forest and Landscape Institute. We also wish to thank our colleagues for useful discussions and input, and the efforts of the two anonymous reviewers for their contributions in improving the quality of the manuscript.

References

[1] Granhus, A., G. Hylen, and J.-E. Ørnelund Nilsen, *Skogen i Norge: Statistikk over skogforhold og skogressurser i Norge registrert i perioden 2005–2009,* in *Ressursoversikt fra Skog og landskap*2012, Norwegian Forest and Landscape Institute. p. 85.

[2] Anon. *The World Bank data on Population Density* 2014 [cited 2014 09.10.2014]; Available from: http://data.worldbank.org/indicator/EN.POP.DNST.

[3] MoE, *Natural benefits - on the values of ecosystems services: Report from an expert commission appointed by the Norwegian Government to the Ministry of the Environment on 29 August 2013.,* in *Official Norwegian Report NOU 2013: 10 Summary*2013, Norwegian Ministry of the Environment Oslo. p. 38.

[4] Daily, G. C., S. Polasky, J. Goldstein, P. M. Kareiva, H. A. Mooney, L. Pejchar, T. H. Ricketts, J. Salzman, and R. Shallenberger, *Ecosystem services in decision making: time to deliver.* Frontiers in Ecology and the Environment, 2009. **7**(1): p. 21–28.

[5] Clarke, L. and K. Jiang, *Chapter 6: Assessing transformation pathways,* in *Climate Change Mitigation - Contribution by Working Group III to the Fifth IPCC Assessment Report.* 2014: Geneva. p. 141.

[6] Luyssaert, S., M. Jammet, P. Stoy, S. Estel, J. Pongratz, E. Ceschia, G. Churkina, A. Don, K.-H. Erb, M. Ferlicoq, B. Gielen, T. Grünwald, R. Houghton, K. Klumpp, A. Knohl, T. Kolb, and et al., *Land management and land-cover change have impacts of similar magnitude on surface temperature.* Nature Climate Change, 2014. **4**(5): p. 389–393.

[7] Lambin, E. F. and P. Meyfroidt, *Global land use change, economic globalization, and the looming land scarcity.* Proceedings of the National Academy of Sciences, 2011. **108**(9): p. 3465–3472.

[8] Kraxner, F., E.-M. Nordström, P. Havlík, M. Gusti, A. Mosnier, S. Frank, H. Valin, S. Fritz, S. Fuss, G. Kindermann, I. McCallum, N. Khabarov, H. Böttcher, L. See, K. Aoki, E. Schmid, L. Máthé, and M. Obersteiner, *Global bioenergy scenarios – Future forest development, land-use implications, and trade-offs.* Biomass and Bioenergy, 2013. **57**(0): p. 86–96.

[9] LMD, *White paper on agricultural and food policy,* 2011, Landbruks-og matdepartement (LMD) [The Ministry of agriculture and food].

[10] Duncker, P. S., S. M. Barreiro, G. M. Hengeveld, T. Lind, W. L. Mason, S. Ambrozy, and H. Spiecker, *Classification of Forest Management Approaches: A New Conceptual Framework and Its Applicability to European Forestry.* Ecology and Society, 2012. **17**(4).

[11] Başkent, E., S. Keleş, A. Kadioğullari, and Ö. Bingöl, *Quantifying the Effects of Forest Management Strategies on the Production of Forest Values: Timber, Carbon, Oxygen, Water, and Soil.* Environmental Modeling & Assessment, 2011. **16**(2): p. 145–152.

[12] Kimmins, J. H., *Balancing act: environmental issues in forestry.* 2011: UBC Press.

[13] Reinhard, S., C. A. Knox Lovell, and G. J. Thijssen, *Environmental efficiency with multiple environmentally detrimental variables; estimated with SFA and DEA.* European Journal of Operational Research, 2000. **121**(2): p. 287–303.

[14] Dios-Palomares, R. and J. M. Martínez-Paz, *Technical, quality and environmental efficiency of the olive oil industry.* Food Policy, 2011. **36**(4): p. 526–534.

[15] Thanh Nguyen, T., V.-N. Hoang, and B. Seo, *Cost and environmental efficiency of rice farms in South Korea.* Agricultural Economics, 2012. **43**(4): p. 369–378.

[16] FSC, *FSC Principles and Criteria for Forest Stewardship, in FSC-STD-01–001 (version 4)*2002, Forest Stewardship Council: Bonn. p. 13.

[17] PEFC, *Living Forests - Standard for sustainable forest management in Norway,* 2006. p. 38.

[18] Anon., *Living Forests: Standard for sustainable forest management in Norway,* 2006: Oslo. p. 38.

[19] Ottaviani, G., B. Talbot, M. Nitteberg, and K. Stampfer, *Workload Benefits of Using a Synthetic Rope Strawline in Cable Yarder Rigging in Norway.* Croatian Journal of Forest Engineering, 2011. **32**(2).

[20] Aalmo, G. O. and B. Talbot, Operator performance improvement through training in a controlled cable yarding study. International Journal of Forest Engineering., 2014 25(1):p. 5–13.

[21] Devoli, G., T. Bargel, A. Taurisano, T. Wiig, H. Berg, E. Øydvin, R. Hermanns, and L. Hansen, *Identifying Needs and Areas for Future Landslide Hazard Mapping in Norway,* in *Landslide Science and Practice,* C. Margottini, P. Canuti, and K. Sassa, Editors. 2013, Springer Berlin Heidelberg. p. 223–230.

[22] Ziesak, M., A. F. Marques, J. Rasinmaki, C. Rosset, K. Nummila, J. Scholz, M. Mittlboeck, J. Pinho de Sousa, J. Häkli, and D. Rommel, *Advances in forestry control and automation systems i Europe - FOCUS: the concept idea in a multinational EU research project,* in *Proceedings of the 6th Precision Forestry Symposium: The anchor of your value chain,* P. Ackerman, E. Gleasure, and H. Ham, Editors. 2014, Faculty of AgriSciences, Stellenbosch University: Stellenbosch, South Africa. p. 114.

[23] Beaudoin, D., J. M. Frayret, and L. LeBel, *Hierarchical forest management with anticipation: an application to tactical–operational planning integration.* Canadian Journal of Forest Research, 2008. **38**(8): p. 2198–2211.

[24] Väätäinen, K., A. Asikainen, L. Sikanen, and A. Ala-Fossi, *The cost effect of forest machine relocations on logging costs in Finland.* Forestry Studies Metsanduslikud Uurimused, 2006. **45:** p. 135–141.

[25] Smaltschinski, T., U. Seeling, and G. Becker, *Clustering forest harvest stands on spatial networks for optimised harvest scheduling.* Annals of Forest Science, 2012. **69**(5): p. 651–657.

[26] Kolka, R., A. Steber, K. Brooks, C. H. Perry, and M. Powers, *Relationships between Soil Compaction and Harvest Season, Soil Texture, and Landscape Position for Aspen Forests.* Northern Journal of Applied Forestry, 2012. **29**(1): p. 21–25.

[27] Borges, P., E. Bergseng, and T. Eid, *Adjacency constraints in forestry – a simulated annealing approach comparing different candidate solution generators.* 2014. Vol. 6. 2014.

[28] Strandgard, M., D. Walsh, and M. Acuna, *Estimating harvester productivity in Pinus radiata plantations using StanForD stem files.* Scandinavian Journal of Forest Research, 2012. **28**(1): p. 73–80.

[29] Suvinen, A. and M. Saarilahti, *Measuring the mobility parameters of forwarders using GPS and CAN bus techniques.* Journal of Terramechanics, 2006. **43**(2): p. 237–252.

[30] Palander, T., Y. Nuutinen, A. Kariniemi, and K. Väätäinen, *Automatic Time Study Method for Recording Work Phase Times of Timber Harvesting.* Forest Science, 2013. **59**(4): p. 472–483.

[31] Næsset, E. and J. G. Gjevestad, *Performance of GPS Precise Point Positioning Under Conifer Forest Canopies.* Photogrammetric Engineering & Remote Sensing, 2008. **74**(5): p. 661–668.

[32] Ringdahl, O., P. Hohnloser, T. Hellström, J. Holmgren, and O. Lindroos, *Enhanced Algorithms for Estimating Tree Trunk Diameter Using 2D Laser Scanner.* Remote Sensing, 2013. **5**(10): p. 4839–4856.

[33] Mousazadeh, H., *A technical review on navigation systems of agricultural autonomous off-road vehicles.* Journal of Terramechanics, 2013. **50**(3): p. 211–232.

[34] Miettinen, M., M. Ohman, A. Visala, and P. Forsman. *Simultaneous Localization and Mapping for Forest Harvesters.* in *Robotics and Automation, 2007 IEEE International Conference on.* 2007.

[35] Wang, J., C. B. LeDoux, and Y. Li, *Simulating Cut-to-Length Harvesting Operations in Appalachian Hardwoods.* International Journal of Forest Engineering, 2005. **16**(2): p. 11–27.

[36] Gerasimov, Y., V. Senkin, and K. Väätäinen, *Productivity of single-grip harvesters in clear-cutting operations in the northern European part of Russia.* European Journal of Forest Research, 2012. **131**(3): p. 647–654.

[37] Barth, A., J. Möller, L. Wilhelmsson, J. Arlinger, R. Hedberg, and U. Söderman, *A Swedish case study on the prediction of detailed product recovery from individual stem profiles based on airborne laser scanning.* Annals of Forest Science, 2014: p. 1–10.

[38] Murphy, G., *Determining Stand Value and Log Product Yields Using Terrestrial Lidar and Optimal Bucking: A Case Study.* Journal of Forestry, 2008. **106**(6): p. 317–324.

[39] Chauhan, S., J. M. Frayret, and L. LeBel, *Supply network planning in the forest supply chain with bucking decisions anticipation.* Annals of Operations Research, 2011. **190**(1): p. 93–115.

[40] Dems, A., L.-M. Rousseau, and J.-M. Frayret, *Effects of different cut-to-length harvesting structures on the economic value of a wood procurement planning problem.* Annals of Operations Research, 2013: p. 1–22.

[41] Manner, J., T. Nordfjell, and O. Lindroos, *Effects of the number of assortments and log concentration on time consumption for forwarding.* Silva Fennica, 2013. **47**(4): p. 19p.

[42] Arce, J. E., C. Carnieri, C. R. Sanquetta, and A. F. Filho, *A Forest-Level Bucking Optimization System that Considers Customer's Demand and Transportation Costs.* Forest Science, 2002. **48**(3): p. 492–503.

[43] Gautam, S., L. LeBel, and D. Beaudoin, *Agility capabilities in wood procurement systems: a literature synthesis.* International Journal of Forest Engineering, 2013. **24**(3): p. 216–232.

[44] Osterkamp, W. R., C. R. Hupp, and M. Stoffel, *The interactions between vegetation and erosion: new directions for research at the interface of ecology and geomorphology.* Earth Surface Processes and Landforms, 2012. **37**(1): p. 23–36.

[45] Roa-García, M. C., S. Brown, H. Schreier, and L. M. Lavkulich, *The role of land use and soils in regulating water flow in small headwater catchments of the Andes.* Water Resources Research, 2011. **47**(5): p. W05510.

[46] Laudon, H., R. Sponseller, R. Lucas, M. Futter, G. Egnell, K. Bishop, A. Ågren, E. Ring, and P. Högberg, *Consequences of More Intensive Forestry for the Sustainable Management of Forest Soils and Waters.* Forests, 2011. **2**(1): p. 243–260.

[47] Sundberg, U. and C. R. Silversides, *Operational efficiency in Forestry.* Forestry Sciences. 1998, Dordrecht: Kluwer Academic Publishers. 219.

[48] Aruga, K., J. Sessions, and E. S. Miyata, *Forest road design with soil sediment evaluation using a high-resolution DEM.* Journal of Forest Research, 2005. **10**(6): p. 471–479.

[49] Contreras, M., P. Aracena, and W. Chung, *Improving Accuracy in Earthwork Volume Estimation for Proposed Forest Roads Using a High-Resolution Digital Elevation Model.* Croatian Journal of Forest Engineering, 2012. **33**(1): p. 125–142.

[50] James, L. A., D. G. Watson, and W. F. Hansen, *Using LiDAR data to map gullies and headwater streams under forest canopy: South Carolina, USA.* CATENA, 2007. **71**(1): p. 132–144.

[51] Perroy, R. L., B. Bookhagen, G. P. Asner, and O. A. Chadwick, *Comparison of gully erosion estimates using airborne and ground-based LiDAR on Santa Cruz Island, California.* Geomorphology, 2010. **118**(3–4): p. 288–300.

[52] Johnsrud, T.-E., J. Bjerketvedt, J. K. Lileng, N. O. Kyllo, and D. Skjølaas, *Normaler for landbruksveier - med byggebeskrivelse [Prescriptions for low volume roads - with construction guidelines]*, 2013, Norwegian Agricultural Authority. p. 82.

[53] Megahan, W. F. and W. J. Kidd, *Effects of logging and logging roads on erosion and sediment deposition from steep terrain* Journal of Forestry 1972. **70**(3): p. 136–141.

[54] Pierzchał,a, M., B. Talbot, and R. Astrup, *Estimating Soil Displacement from Timber Extraction Trails in Steep Terrain: Application of an Unmanned Aircraft for 3D Modelling.* Forests, 2014. **5**(6): p. 1212–1223.

[55] Smidt, M. F. and R. K. Kolka, *Alternative skid trail retirement options for steep terrain logging, in The 24th Annual Meeting of The Council of Forest Engineering (COFE). Appalachian Hardwoods: Managing Change,* J. Wang, M. Wolford, and J. F. McNeel, Editors. 2001, USDA Forest Service: Snowshoe, West Virginia. p. 6.

[56] Christopher, E., A and R. Visser, *Methodology for evaluatingpost harvest erosion risk for the protection of water quality.* New Zealand Journal of Forestry, 2007: p. 20–25.

[57] Sawyers, B. C., M. C. Bolding, W. M. Aust, and W. A. Lakel, *Effectiveness and implementation costs of overland skid trail closure techniques in the Virginia Piedmont.* Journal of Soil and Water Conservation, 2012. **67**(4): p. 300–310.

[58] Søvde, N. E., A. Løkketangen, and B. Talbot, *Applicability of the GRASP metaheuristic method in designing machine trail layout.* Forest Science and Technology, 2013. **9**(4): p. 187–194.

[59] SSB. *Focus on Forests.* 10.04.20 2012 [cited 2012 30.06]; Available from: http://www.ssb.no/10/04/20/skog/.

[60] Egglishaw, H., R. Gardiner, and J. Foster, *Salmon catch decline and forestry in Scotland.* Scottish Geographical Magazine, 1986. **102**(1): p. 57–61.

[61] Contreras, M. and W. Chung, *A computer approach to finding an optimal log landing location and analyzing influencing factors for ground-based timber harvesting.* Canadian Journal of Forest Research, 2007. **37**(2): p. 276–292.

[62] Sakai, H., T. Nordfjell, K. Suadicani, B. Talbot, and E. Bøllehuus, *Soil compaction on forest soils from different kinds of tires and tracks and possibility of accurate estimate.* Croatian Journal of Forest Engineering, 2008. **29**(1): p. 15–27.

[63] Timmermann, V. and J. Dibdiakova, *Greenhouse gas emissions from forestry in East Norway.* The International Journal of Life Cycle Assessment, 2014. **19**(9): p. 1593–1606.

[64] Søvde, N. E., B. Talbot, J. Bjerketvedt, and M. Pierzchala, *Improving parameter estimates in modelling the influence of pitch and roll on*

forwarder driving speed, in Proceedings of the 6th Precision Forestry Symposium: The anchor of your value chain, P. Ackerman, E. Gleasure, and H. Ham, Editors. 2014, Faculty of AgriSciences, Stellenbosch University: Stellenbosch, South Africa.

[65] Ringdahl, O., T. Hellström, I. Wästerlund, and O. Lindroos, *Estimating wheel slip for a forest machine using RTK-DGPS*. Journal of Terramechanics, 2012. **49**(5): p. 271–279.

[66] Eliasson, L., *Effects of forwarder tyre pressure on rut formation and soil compaction*. Silva Fennica, 2005. **39**(4): p. 549–557.

[67] Edlund, J., E. Keramati, and M. Servin, *A long-tracked bogie design for forestry machines on soft and rough terrain*. Journal of Terramechanics, 2013. **50**(2): p. 73–83.

[68] Horn, R., J. Vossbrink, and S. Becker, *Modern forestry vehicles and their impacts on soil physical properties*. Soil and Tillage Research, 2004. **79**(2): p. 207–219.

[69] Purfürst, F. T. and J. Erler, *The Human Influence on Productivity in Harvester Operations*. International Journal of Forest Engineering, 2011. **22**(2): p. 15–22.

[70] Purfürst, F. T., *Learning curves of harvester operators*. Croatian Journal of Forest Engineering, 2010. **31**(2): p. 89–97.

[71] Lindroos, O., *Scrutinizing the Theory of Comparative Time Studies with Operator as Block Effect*. International Journal of Forest Engineering, 2010. **21**(1): p. 10.

[72] Risbøl, O., C. Briese, M. Doneus, and A. Nesbakken, *Monitoring cultural heritage by comparing DEMs derived from historical aerial photographs and airborne laser scanning*. Journal of Cultural Heritage, 2014.

[73] Nielsen, A. B., E. Heyman, and G. Richnau, *Liked, disliked and unseen forest attributes: Relation to modes of viewing and cognitive constructs*. Journal of Environmental Management, 2012. **113**(0): p. 456–466.

[74] Gundersen, V. S. and L. H. Frivold, *Public preferences for forest structures: A review of quantitative surveys from Finland, Norway and Sweden*. Urban Forestry & Urban Greening, 2008. **7**(4): p. 241–258.

[75] Laband, D. N., *The neglected stepchildren of forest-based ecosystem services: Cultural, spiritual, and aesthetic values*. Forest Policy and Economics, 2013. **35**(0): p. 39–44.

[76] Schei, F. H., H. H. Blom, I. Gjerde, J.-A. Grytnes, E. Heegaard, and M. Sætersdal, *Conservation of epiphytes: Single large or several small host trees?* Biological Conservation, 2013. **168**(0): p. 144–151.

[77] Tangeland, T., Ø. Aas, and A. Odden, *The Socio-Demographic Influence on Participation in Outdoor Recreation Activities – Implications for the Norwegian Domestic Market for Nature-Based Tourism.* Scandinavian Journal of Hospitality and Tourism, 2013. **13**(3): p. 190–207.

[78] Sievänen, T., D. Edwards, P. Fredman, F. Søndergaard Jensen, and O. I. Vistad, *Indicators for sustainable recreational use of forests and other natural resources - experiences from Northern Europe,* in *Protected Areas and Place Making: How do we provide conservation, landscape management, tourism, human health and regional development?,* T. C. Magro, et al., Editors. 2013, Forestry Sciences Departament – ESALQ/USP: Foz do Iguacu, Brazil. p. 204.

[79] Pasalodos-Tato, M., A. Mäkinen, J. Garcia-Gonzalo, J. G. Borges, T. Lämås, and L. O. Eriksson, *Review. Assessing uncertainty and risk in forest planning and decision support systems: review of classical methods and introduction of new approaches.* 2013, 2013. **22**(2).

Biographies

Bruce Talbot is researcher in forest operations and works with a wide range of related issues in biomass procurement and timber harvesting and supply logistics. More recently, his research work has centered around steep terrain logging in Norway focusing both on the productivity and the environmental impact of these harvesting systems.

Rasmus Astrup is researcher and Head of the Department of forest resources. He works across multiple disciplines although his main research interest is in forest inventory and forest modelling with a special emphasis on developing new approaches for quantification of various aspects of forestry. Within forest inventory his main interest is in the development of efficient and robust methods combining remote sensing with field-based inventories.

A Process Perspective on the Timber Transport Vehicle Routing Problem

Jonas Lindström[1] and Dag Fjeld[2]

[1] Södra Skog, Billingsfors, Sweden
[2] Faculty of Forestry, Swedish University of Agricultural Sciences, Umeå, Sweden
Corresponding Authors: Jonas.Lindstrom@sodra.com; Dag.Fjeld@slu.se

Received 2 June 2014; Accepted 27 November 2014;
Publication 19 March 2015

Abstract

The aim of this study was to map how the timber transport vehicle routing problem was solved in practice and which consequences different ways of solving the problem had for service and efficiency. A process perspective was employed for the mapping and the ways of solving the routing problem were expressed in terms of a series of planning and control activities. Fifteen haulage contractors from the Södra Skogägarna forest owners association were selected for the mapping. The mapping resulted in a basic process model and 2 main variants. Key performance indicators for both service and economic efficiency were collected for a one-year period. The contractors' service levels to suppliers were measured by the proportion of transport orders completed within a specified period. The contractors' economic efficiency was measured by their net operating margin. The results show that contractor net operating margins decreased (from 15% to 1%) with increasing levels of supplier service (from 89.5 to 97% of orders completed within 5 weeks). Within this gradient, those using the complete process model had an average net operating margin of 4.1%. Those using a simplified model (with fewer service restrictions) had an average margin of 9.2%.

Keywords: haulage contractors, truck routing, service levels, economic efficiency.

Journal of Green Engineering, Vol. 4, 291–306.
doi: 10.13052/jge1904-4720.443

Introduction

In Sweden, truck transport of roundwood is estimated to account for an annual emission of 428 000 tons CO_2 (Andersson & Frisk 2013). These emissions are in direct proportion to logging truck fuel consumption. Studies of logging truck fuel consumption (Svenson & Fjeld 2012) show that gross vehicle weight and road grade are the main influencing factors. Given that gross vehicle weight for each truck load unavoidably varies between the empty (tare) weight of approx. 20 tons and fully loaded weight of 60 tons, development efforts to reduce emissions should be focused on reducing empty travel, choosing suitable routes for the loaded transport from forest to mill and further reducing tare weight.

Vehicle routing is often described as having the goal of maximizing capacity utilization or minimizing total costs for a given service (Haksever et al. 2000). In roundwood trucking, with its potentially high proportion of unloaded transport from the mill back to the forest, routing studies are often focused on the reduction of the unloaded distance through backhauling. Some companies have introduced software for vehicle routing and scheduling (Weintraub et al. 1996, Savola et al 2004). These solutions have gradually been adapted to capture the restrictions handled by haulage contractors in practice (Karanta et al. 2000). Few studies, however, have been published documenting how these contractors solve their routing problems in practice.

Swedish transport management is an integral part of wood supply management. The high proportion of wood from non-industrial private forest owners and distribution of saw and pulp mills requires the coordination of a number of assortments from scattered harvesting sites to multiple mills. These operations are typically managed by the supply organization (transport service buyer) where central administration contracts transport capacity and then distribute periodic-specific transport goals to regional transport managers and their respective contractors (transport service providers). At the operational level vehicle routing becomes the responsibility of the service provider. The supply organization, however normally retains close control over truck deliveries as a possibility for compensating for disturbances in the other parts of the transport system. For a haulage contractor the goal is then to maximize the economic result, within the restrictions of the agreed service level. Payment is done according to a tariff formula calculating the price per transported unit based on a fixed rate per unit and variable rate per unit. The fixed rate per unit is intended to cover the indirect transport work (loading/unloading) and the variable rate is intended to cover the direct transport work given the loaded distance from mill to forest and an accepted proportion of loaded and unloaded driving.

Much literature is found recommending process improvement both within and between organizations for the purpose of increasing income and reducing costs, however, only a few forest-sector studies have been done on this topic. Mäkinen (1993, 2001) and Soirinsu and Mäkinen (2009) examine transport contractor profitability from a business strategy perspective. Erlandsson (2008) examines contractor profitability in terms of the task environment which also includes some interfaces with service buyers, but without examining process configuration as an influencing factor.

Aims of This Study

The first aim of this study is to map contractor-level processes for routing of self-loading logging trucks. The second aim is to identify main variants of the process model and see if these differences are linked to service and economic efficiency levels.

Methods

The study was done in two parts. The first was process mapping and the second was the search for links to service and economic efficiency levels. The process mapping started with personal interviews and the formation of draft processes for individual contractors. When the draft maps for individual contractors were ready the search began for common features linking the different drafts to a general model. The main variants of the general model were then defined and the corresponding service and profitability levels were compared.

Process Mapping

The study was hosted by Södra Skogsägarna, a forest owners association in south Sweden. Multi-truck contractors were randomly selected from each of the organization's 3 regions (East, West, South). The distribution of contractors per region was 6 in the East, 6 in the West and 3 in the South (15 in total). Each of the respondents were contacted first by mail (to explain the aim of the study) and later by telephone (to book time and place for the interview). Participation was agreed to under conditions of anonymity.

The process mapping was based on the methods and nomenclature described by Larsson and Ljungberg (2001) who specify a variant of mapping called design-process where complex structures must be formulated from semantic descriptions in the absence of physically observable activities. Larsson and Ljungberg (2001) refer to three levels of detail: process, sub-process and activity. Within this hierarchy any process is assumed to include a number of components including input (which triggers the start of the process/sub-process or activity), activity (which transforms the input), resources (which are needed to do the activity), information (which supports or controls the activity) and output (which is the result of the activity and may be the input for the next activity).

The personal interviews with each contractor also covered a number of themes other than the explicit mapping of the routing process. These included descriptive information on the contractor's enterprise, their cooperation with the service buyer's transport managers and other parameters influencing the contractor's task environment. The interviewer asked direct and simple questions according to a pre-prepared structure allowing the respondent complete freedom to formulate complex answers. The interviewer used a series of empty process diagrams to help formulate the process during the interview. Interviews were recorded on a Dictaphone for future reference. After the completion of the 15 interviews, each contractor's routing activities were defined and named. The draft process maps which were filled in during the interviews were then compared to the recorded protocol and adjusted if required. After this the activities were categorized into sub-processes according to similarities in purpose and level of detail. After this the sub-processes and activities were defined and named.

Key Performance Indicators

Key performance indicators were collected for both service and efficiency. The chosen service level indicator was the proportion of assigned transport orders completed (all volumes for assortments delivered) within 5 weeks of initiation. This variable is therefore a measure of service offered to suppliers (forest owners with a delivery contract to Södra Skogsägarna). Average values were taken from the service buyers database (based on input from SDC, the central database for Swedish wood transactions). Data was missing for one contractor. The chosen indicator for contractor economic efficiency was net operating margin defined as the net operating surplus (after financial costs) as a proportion of annual turnover. Net operating margin is a relative term and therefore robust when comparing enterprises of different sizes. Values were available for limited stock companies through the Swedish national database. This data was not available for 4 contractors which had other forms of ownership. The analysis of how enterprise-level service and profitability corresponds with process configuration was done quite simply. Average values of contractor service and profitability were compared between the variants of process configuration. Scatter plots between variables from individual contractors were used to visualize eventual relationships.

Results

The contractors in the study had between 2 and 12 trucks per enterprise and delivered wood to between 5 and 15 mills. Each truck delivered approx. 40 000 m³/yr with a typical utilization of 4500 hrs/yr (Table 1).

The average annual turnover per contractor in the study was approx. 12 million SEK. The average net operating margin (profit before financial costs as a proportion of annual turnover) was 5% but varied from -3% to 15%. The average service level (% of transport orders completed within 5 weeks) was 93% but varied from 84 to 97% (Table 2).

The Contractor Routing Process

After a comparison of all the individual contractor models, a basic model was formed consisting of all the activities which the majority of contractors (8 of 15) used to solve their own routing problems. These activities were

Table 1 Descriptive statistics for operating context (no. of trucks and no. of mills being serviced, typical delivery distances) and truck utilization (m³, km and hrs per yr and truck) for the contractors in the study

	Mean	Median	Max	Min	N
trucks/contractor	4,87	4	12	2	15
mills/contractor	9,4	8,5	15	5	14
delivery distance (km)	80	80	120	50	15
m³/yr/truck	47606	39743	100000	32000	14
km/yr/truck	185890	180000	230000	135000	15
hrs/yr/truck	4525	4500	5405	4000	12

Table 2 Descriptive statistics for some economic indicators (turnover, profit, net annual margin) and supplier service parameters (% of transport orders completed within 5 weeks) for the contractors in the study

	Mean	Median	Max	Min	N
Net annual turnover (1000 SEK)	12 606	12 873	19434	6309	11
Annual Profit (1000 SEK)	295	206	1123	2	11
Net annual margin (%)	5%	3%	15%	–3%	11
Supplier service level (%)	93,2	94,3	96,9	84,1	14

aggregated into the 4 sub-processes (Figure 1). These and their respective activities are described below.

Information gathering (1). This sub-process consists of one activity. Receipt of new transport orders (1.1) is a daily activity where the contractor receives new transport orders (delivery responsibility for a specific harvesting site) from the forest owners association. This occurs, either through direct contact with the service buyers transport manager or by downloading the new assignments directly from the service buyer's information system. After this sub-process the contractor has a complete list over his transport orders (all harvesting sites where he is responsible).

Preparatory planning (2) consists of 4 activities. This sub-process gives the contractor an overview of the restrictions, priorities and possibilities for vehicle routing. Mill quota follow-up (2.1) is for tracking the contractor's weekly quota for volumes per assortment to de delivered to the respective mills. This activity monitors the volumes delivered so far and how much is left for the days remaining. This activity can also include increases or decreases in the quota if supply or demand conditions require so. Ranking of transport orders (2.2) is when the contractor ranks all the transport orders based on a selection of priority factors. Unless special conditions exists the default priority is based on the date the transport order was assigned (oldest first). Analysis of geographic flow patterns (2.3) is when the contractor examines the patterns of wood flow to see where there exists potential for backhauling. Contact with other contractors (2.4) is for when a suitable wood flow pattern for backhauling exists between contractor transport orders and the contractor makes contact with another contractor to request an exchange of volumes to realize the backhaul. After this sub-process the contractor has a ranked list of transport orders indicating the sequence they should be delivered to meet both supplier and mill service requirements while reducing the proportion of unloaded driving.

Problem solution (3). This sub-process consists of 3 activities. These activities determine how the vehicle routing will be done during the planning period in question. Filtering of infeasible transport orders (3.1) is when the contractor filters out harvesting sites which are temporarily unavailable due to weather conditions or limited opening hours for wood receival at the mill. Clustering of small volumes into whole loads (3.2) is when the contractor locates smaller volumes (of the same mill destination) within acceptable distances to aggregate into whole loads. Search for load sequences (3.2) is when the contractor factors in the working hours of the individual operator and combine loads into sequences that give the operators full shifts that conclude

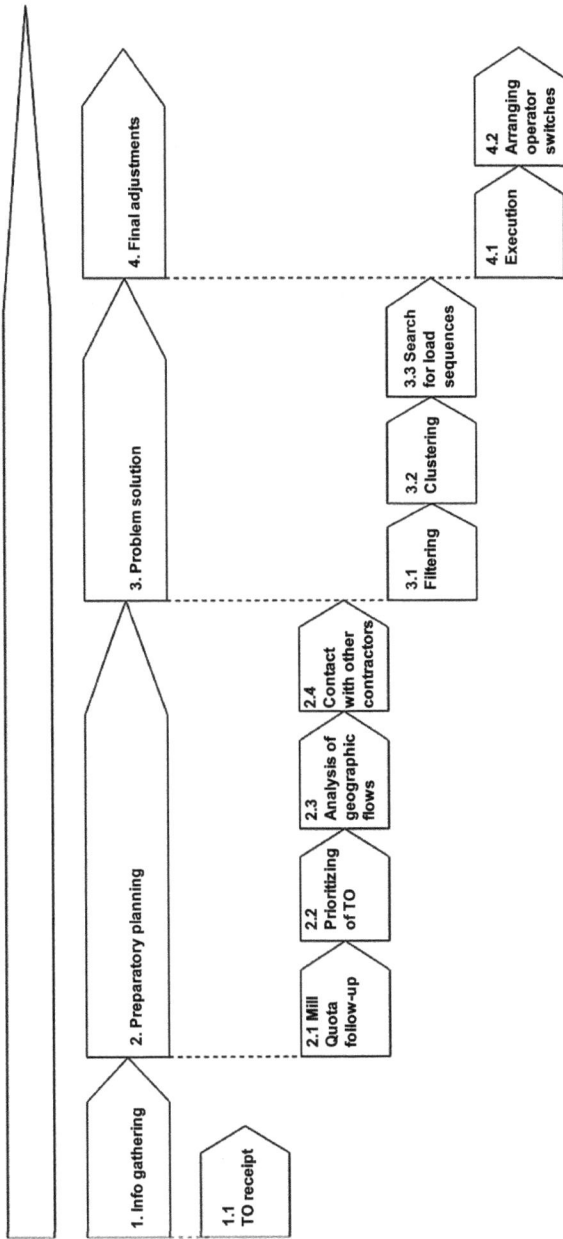

Figure 1   A basic process model for routing of self-loading logging trucks with 4 sub- processes and 10 activities. The model consists of the activities that the majority of haulage contractors used in their vehicle routing.

close to their home bases. After this sub-process the contractor has solved the daily routing for his trucks.

Final adjustments (4) – consists of 2 activities. Execution (4.1) is the operators' execution of the individual delivery and detailed planning of the path to each harvesting site for loading. Arranging operator switches (4.2) is when the operators contact each other and agree to an exact meeting place for changing operators between shifts. After this sub-process the contractor's routing solution has been executed and operator schedules are coordinated in detail.

Key Performance Indicators for Variants of the Routing Process

Seven contractors of the 15 studied used another variant of the basic model than described in Figure 1. These had a simplified preparatory planning sub-process (2) where follow-up of mill quotas (2.1) was not taken into consideration. Two contractors of the 15 studied used a simplification of the problem solution sub-process (3) where clustering of small volumes (3.2) and the explicit search for optimal load sequences (3.3) was not included.

Figure 2 shows that for those contractors working with either the basic model or the variant with simplified preparatory planning (2) profitability decreased (from 15% to 1% net operating margin) with increasing levels of supplier service (from 89.5 to 97% of orders completed within 5 weeks).

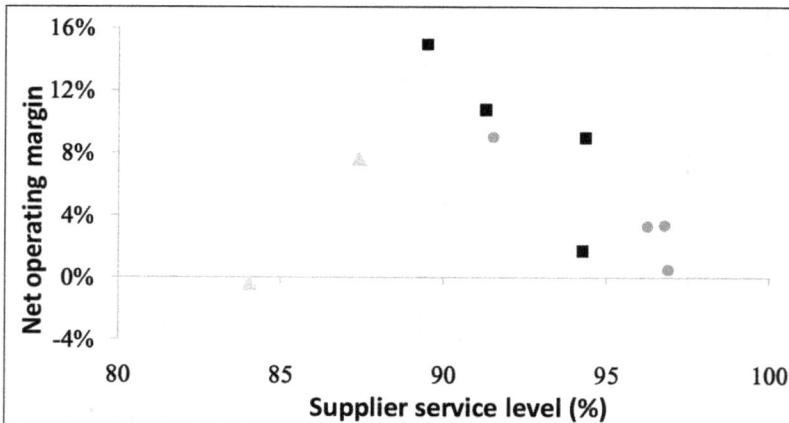

Figure 2 Scatter plot between net operating margin and supplier service levels for 10 contractors (circles = contractors with the complete basic process model, squares = contractors with simplified preparatory planning, triangles = contractors that have a simplified solution sub-process.

Within this gradient, those using the complete basic process model had an average net operating margin of 4.1% while those not limited by quota follow-up (2.1) had an average margin of 9.2%. Those contractors working with the complete model had higher supplier service levels in all three regions (Figure 3). Figure 4 shows that profitability decreased with increasing annual operating hours per truck, regardless of which process model the contractor

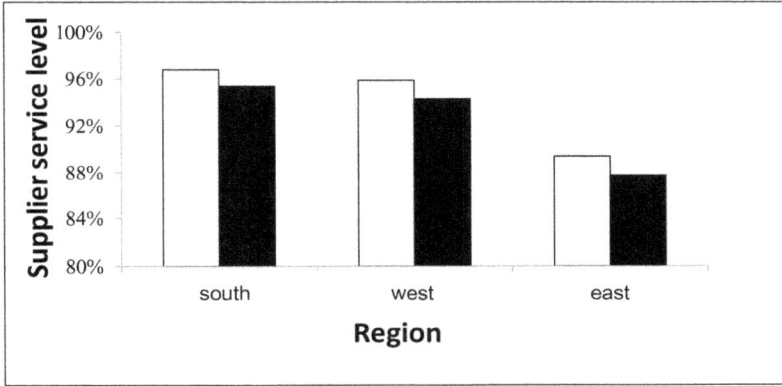

Figure 3 Mean supplier service levels for haulage contractors in each region grouped into whether they had a simplified preparatory planning sub-process or not (black columns = simplified sub-process, white columns = complete sub-process).

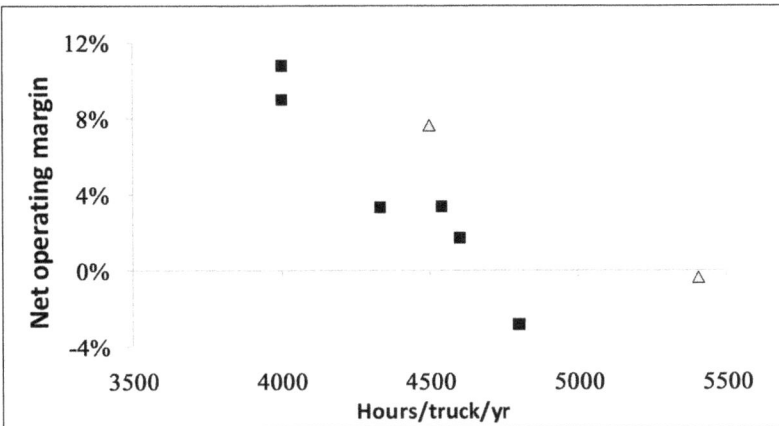

Figure 4 Scatter plot between contractors' net operating margin and the number of annual operating hours per trucks (squares = contractors with a complete problem solution sub-process, triangles = contractors with a simplified problem solution sub-process).

used. However, the contractors with a simplified problem solution sub-process (3) had a higher number of operating hours than other contractors for the same level of profitability.

Discussion

Earlier studies of logging truck routing have been mainly focused on developing mathematical models to give optimal- or near-optimal solutions. In contrast, this study shows how haulage contractors do their vehicle routing in practice. Three main variants were found: a complete process model including all sub-processes/activities, and two variants missing either key activities in the sub-process *preparatory planning* (activity quota follow-up) or the sub-process *problem solution* (activity search for load sequences). Compared to the complete model, the variant with the fewest restrictions (missing quota follow-up) was associated with a higher net operating margin and lower supply service levels. The variant with the least focus on the actual routing (missing an active search for the best load sequences) was associated with a higher number of operating hours per year for the same net operating margin. Regardless of the process model used by the individual contractor, a plot of all observations showed an overall trend of lower net operating margins for contractors with high supplier service levels (% of orders completed within 5 weeks). In practice, higher service levels for mills and supplier service levels represent increasingly tighter restrictions for the contractors' routing. These issues are explored in more detail below.

Karanta et al. (2000) characterizes the logging truck routing problem (otherwise known as the TTVRP) as one of the most difficult problems to solve within the world of operations research (OR). Given this, it seems a paradox that roundwood trucking functions as well as it does. The main difference between mathematical formulations and transport management in practice, however, is the division of large problems/systems into smaller sub-problems/systems (decentralization). While this practice poses the risk of sub-optimization, it makes the problems possible to solve manually without advanced decision support systems. Karanta (2000) mentions two particular challenges of solving this problem with OR methods; first, an unusually high number of constraints which must be taken into consideration and second, the difficulty of specifying a general formulation for the goal function. Regarding the first challenge (number of constraints), problem solving theory would interpret the preparatory planning sub-process as focusing on the most critical factors of the task environment. In this respect the basic process model

(Figure 1) excludes certain potential wood flows already at the outset and enables focus on the most critical constraints. Regarding the second challenge (formulating the goal function), a typical difficulty is conflicting perspectives between different parts of the system (e.g. supplier vs. contractor. vs. mill). In this respect, the preparatory planning sub-process can be interpreted as handling potentially opposing goals between suppliers and mills before beginning to address specific sequences/solutions.

The sequence of activities used by the contractors in this study effectively strips down the problem space to its most critical areas. In an extreme case, a few priority mills might require immediate deliveries which are only possible to fill from a few landings where the critical latest date is near. This may limit the number of potential pick-up and delivery points for each truck within the contractor's "home territory" to a relatively short list. In this case, a contractor could arrange an acceptable sequence of loads for the day using only knowledge of average trip times, making the routing problem much easier to solve (Gerasimov et al. 2013). With respect to reducing emissions, however, the need to further exploit backhauling possibilities cannot be ignored. Given free reign, contractors create efficient routes based on stable wood flows. However, how close they can come to the most efficient routes is determined by the preparatory-planning (2) where potentially efficient combinations of landings or mills can be removed from the priority list. Similar removals occur during filtering (3.1) of the problem solution (3) where, for example, mill opening hours may further shorten the hauler's list of feasible deliveries. A typical example of this is when only one of the two geographically opposing flows of a backhaul can be delivered during a second evening shift. Empirical support for this is shown by Erlandsson (2008) with a correlation between contractor operating margins and the proportion of deliveries to 24-hour open mills.

One reason that responsibility for backhaul planning is most often delegated to the contractor level is that a single contractor may work for a number of competing wood supply organizations. The resulting collaboration between contractors to locate potential backhaul flows has been the subject of a number of studies. Audy et al. (2010) mapped alternative collaboration networks. Frisk (2002) examined decision support systems for helping to locate potential backhaul flows and Karlsson et al. (2006) mapped the inter-organizational processes for arranging backhauling between contractors. The operational feasibility of realizing these backhaul potentials have been examined in both simulation studies (Fjeld 2012) and empirical studies (Auselius 2009) where the levels of roadside stocks were shown to be one of the most critical aspects for high efficiency. It is in this context, when backhauling feasibility becomes dependent on sequences of deliveries, that manual routing begins to resemble an advanced board game. Seen in a game theory perspective, the typical contractors routing process is about moving through a problem space towards a solution which places the whole system in the goal condition. At each step the solver can choose an operator (decision rule) and apply it to get to the next step (state of knowledge) which is hopefully closer to the ultimate goal (in this case, economic efficiency). The way of working mapped in this study has been interpreted as a process which extracts information about the structure of the task environment and uses this for highly selective heuristic search of the problem space for solutions. The particular advantage of the process is that it simplifies a potentially complex problem. The disadvantage is that it may restrict the problem space too much, reducing the degrees of freedom necessary to find the best sequences of pick-ups and deliveries. In this context, both mill quotas and supplier service limits have the same effect – they reduce the number of "legal moves" (degrees of freedom) which can be tested on the way to the next step through the problem space. As a result, the combination of high service demands and a simplified search and solution sub-process clearly reduces the possibility to reach the highest level of utilization and efficiency. In this study the two respondents with the simplified solution sub-process (3.3) worked more hours than the other contractors for the same level of economic efficiency (Figure 4) and this may be consistent with the logic discussed above. At the same time, however, there may other explanations for the absence of explicit solution sub-process (3.3) such as a) this decision may be made by others (higher up in the hierarchy) b) there are few degrees of freedom in the task environment due to other factors or c) the contractor does not have the capacity for further information processing. In general the trend for poorer profitability at high utilization has also been reported by Mäkinen (2001) and has been commonly observed among transport managers. If the above conditions a, b or c are present, increasing hours or volumes cannot compensate for poor planning and may even make the situation worse. So, even though the results can be supported by the logic and earlier studies discussed above, the small sample limits the validity of the results to the context studied.

Final Comments

This study has used a process-perspective to map how the timber transport vehicle routing problem is solved in practice. Although the number of contractors participating in the study was limited, the process perspective was found to be a suitable approach for identifying state-of-the-art in vehicle routing, illustrating main variants as well as their associated service and efficiency levels.

Literature Cited

[1] Andersson, G. and Frisk, M. (2013). Skogbrukets transporter 2010. Arbetsrapport från Skogforsk nr. 791–2013.

[2] Audy, J.-F., D'Amours, S., Lehoux, N., Rönnqvist, M. (2010). Coordination in collaborative logistics. Extended abstract from the International workshop on supply chain management, Brussels, 21–22 January 2010.

[3] Andersson, G., Flisberg, P., Lidén, B., Rönnqvist, M. (2008). RuttOpt – a decision support system for routing of logging trucks. Canadian Journal of Forest Research 38: 1784–1796.

[4] Auselius, J. (2009). Realization of backhauls in roundwood truck transportation – obstacles, possibilities, profits and profit-sharing. Master thesis no. 262, 2009. Swedish University of Agricultural Sciences, Dept. of Forest Resource Management.

[5] Carlsson, D. and Rönnqvist, M. (1998). Tactical planning of forestry transportation with respect to backhauling. Lith-MAT-R-1998-13.

[6] Erlandsson, E. (2008). Success factors for roundwood truck transport companies in Mid-Sweden. Master thesis no. 230, 2008. Swedish University of Agricultural Sciences, Dept. of Forest Resource Management.

[7] Fjeld, D. (2012). Discrete-event simulation of collaboration between transport organizations. In: Lazdiòa, D Jansons, A, Konstantinova I (eds) Forest operations research in the Nordic Baltic region. Proceedings of the 2012 OSCAR conference, Riga, Latvia, October 24–26, 2012. Mežzinātne 25(58)2012.

[8] Gerasimov, J., Sokolov, A. and Fjeld, D. (2013). Improving CTL operations management in Russian logging companies using a new decision support system. Baltic Forestry 19(1): 89–105.

[9] Haksever, C., Render, B., Russel, R. Murdick, R. (2000). Vehicle routing and scheduling. Service management and operations. Prentice-Hall.

[10] Karanta, I., Jokinen, O., Mikkola, T., Savola, J., Bounsaythip, C. (2000). Requirements for a vehicle routing and scheduling system in timber transport. In: Sjöström, K. (ed) 2000. Proceedings from Symposium on Logistics in the forest sector. Timber Logistics Club.

[11] Karlsson, M., Landström, A., Fjeld, D. (2006). Developing the backhaul exchange process between contractors in North Sweden. Forestry Studies 45: 23–32.

[12] Larsson, E. and Ljungberg, A. (2001). Processbaserad verksamhet-sutveckling (Process-based business development). Studentlitteratur (Swedish only).

[13] Mäkinen, P. (2001). Competitive strategies applied by Finnish timber carriers following deregulation. Silva Fennica 35(3): 341–353.

[14] Nilsson, B. (2004). An analysis of roundwood transport planning in Swedish forestry. Master thesis no. 70, 2004. Swedish University of Agricultural Sciences, Dept. of Forest Resource Management.

[15] Soirinsuo, J. and Mäkinen, P. (2009). Growth and economies of scale among timber haulage companies. Journal of Small Business and Enterprise Development Vol. 18. No. 1: 170–184.

[16] Svenson, G. and Fjeld, D. (2012). The influence of road characteristics on fuel consumption for logging trucks. Proceedings from the 2012 International symposium on heavy vehicle transport technology. 16–19 Sept. Stockholm.

[17] Weintraub, A., Epstein, R., Morales, R., Seron, J., Traverso, P. (1996). A truck scheduling system improves efficiency in the forest industries. Interfaces 26: 9–25.

Biographies

Jonas Lindström After finishing his MSc in forestry at the Swedish University of Agricultural Sciences Jonas Lindström worked as the manager of Södra Odlarna at Falkenberg, Sweden and is now harvesting production manager at Södra Skog in Billingsfors, Sweden.

Dag Fjeld After finishing his BSc in forestry at the University of British Columbia and PhD in forest operations at the Norwegian University of Agricultural Sciences Dag Fjeld was a lecturer and associate professor in forest operations at the Swedish University of Agricultural Sciences. He is now working as senior researcher in forest operations at the Norwegian Institute for Forest and Landscape and Professor of forest operations at the Norwegian University of Life Sciences.

Implementing Green Infrastructure and Ecological Networks in Europe: Lessons Learned and Future Perspectives

K. Čivić[1] and L. M. Jones-Walters[2]

[1]*ECNC-European Centre for Nature Conservation, Reitseplein 3, PO Box 90154, 5000LG Tilburg, the Netherlands*
[2]*Alterra, Wageningen UR, Wageningen Campus, 6700 AA Wageningen, the Netherlands*
Corresponding Authors: civic@ecnc.org; lawrence.jones-walters@wur.nl

Received 4 July 2014; Accepted 27 November 2014;
Publication 19 March 2015

Abstract

The impact of landscape fragmentation is well recognised as one of the key contributors to the past and present decline in European wildlife. Ecological networks were seen as a solution to this problem and have been the subject of research, policy and practice for nearly 40 years; resulting in many examples of best practice and lessons learned. More recently the European Commission has introduced the concept of Green Infrastructure (GI) which retains the frame work of ecological networks at its core but which offers a more sophisticated integration of economic and social factors and the delivery of a range of ecosystem services. GI has already been included as a concept in EU strategy and offers much for future policy making and delivery of sectoral integration. The views of stakeholders indicate that there a number of key areas for improvement but confirm the potential of the concept. Further work should consider the practicalities of the full translation of the protected area networks into functional ecological networks and making them integral building blocks of the green infrastructure both at the level of policy and practice. In addition information about how to create actual ecological

Journal of Green Engineering, Vol. 4, 307–324.
doi: 10.13052/jge1904-4720.444

networks at the delivery level, particularly where this has involved stakeholder and public participation needs to be researched and made widely available. The issue of communication; specifically to politicians and decision makers within key sectors (such as spatial planning, transport, industry, etc.) but more widely to researchers, conservation practitioners, businesses and the interested public remain to be fully addressed. [1]

Keywords: ecological networks, green infrastructure, ecosystem services, climate change, resource efficiency, innovation.

1 Introduction

Landscape fragmentation by human activities and infrastructure is a major cause of the well recorded decrease in many European wildlife populations. The current trend of steadily increasing landscape fragmentation contradicts the principle of sustainability and there is a clear and urgent need for action. The continuing proliferation of urban development and transport infrastructure is likely to cause a significant increase in the already existing problems. This increase is likely to be significant, not least because many of the ecological effects of the current levels of fragmentation have yet to manifest fully [2].

Ecological networks represent a very effective tool for combating the effects of fragmentation by: counteracting fragmentation; conserving and buffering core areas; maintaining and establishing ecological connectivity; being a tool for ecological design and planning; being a tool for interaction with other land uses; being an important political instrument. A lot of work has been done in relation to implementing the ecological networks in Europe at various levels with the Pan European Ecological Network (PEEN) as an umbrella initiative. The PEEN approach was successful in reaching its goal of promoting the idea of a pan-European vision of biodiversity conservation through a European ecological network [3]. It is a genuine framework for strategic cooperation and a useful tool for international cooperation, providing all European countries with a single and flexible monitoring and coordination mechanism [4]. In addition, under Article 10 of the Habitats Directive [5], which is one of the cornerstones of European nature legislation, provides "a specific contribution to a coherent European ecological network of protected sites" by designating Special Areas of Conservation (SACs) for habitats listed on Annex I and for species listed on Annex II. These measures are also to be applied to Special Protection Areas (SPAs) classified under Article 4 of the Birds Directive [6]. Currently, ecological networks are taken up as integral parts of Green Infrastructure (GI)– an approach and a new policy tool combining nature conservation and sustainable development promoted by the European Commission [7].

Ecological connectivity remains a priority for international biodiversity conservation policy; illustrated by Target 11 of the Aichi Biodiversity Targets, signed at COP 10 of the Convention on Biological Diversity (CBD) in 2010, which states that: "By 2020, at least 17 per cent of terrestrial and inland water areas and 10 per cent of coastal and marine areas, especially areas of particular importance for biodiversity and ecosystem services, are conserved through effectively and equitably managed, ecologically representative and well-connected systems of protected areas and other effective area-based conservation measures, and integrated into the wider landscape and seascape" [8].

As a response to the Aichi targets, (and therefore included in the European Commission's EU 2020 European biodiversity headline target and 2050 vision whose aim is to halt and reverse the loss of biodiversity across the EU territory of the member states) [9], the GI initiative has relatively recently been launched as a new concept, to coincide with the adoption by the Commission of the GI Strategy, and has the potential to become an important policy instrument [10]. This has been reiterated and further reinforced by the 'Pan-European 2020 Strategy for Biodiversity' adopted by the Council for the Pan-European Biological and Landscape Diversity Strategy (PEBLDS), for the wider Pan-European region [11].

In the context of these developments it is important to reflect on what have we learned from the work carried out previously (over the past four decades) on the development and implementation of ecological networks at various levels (from the pan-European, supra-national, national, regional and local) across Europe and elsewhere; how to best utilise and put into practice the lessons learned; and how to ensure that the GI initiative takes a necessary step beyond what has already been achieved.

This paper provides a summary of the achievements of work that has taken place in Europe over the previous 40 years to establish ecological networks, primarily to combat fragmentation but also to deliver a range of other benefits to biodiversity at different geopolitical and spatial levels. In addition it places this work in the perspective of emerging EU policy on GI and considers how the concept has now been broadened to better reflect the integration of economic and social factors and the delivery of a range of ecosystem services.

2 Ecological Networks – Lessons Learned

The concept of ecological networks is not new; the model has been developed over the past 35–40 years (beginning in the 1970s and 1980s in countries where a strong land use planning tradition had created the institutional environment for allocating functions at the landscape scale) in the context of increasingly fragmented European landscapes. The concept represents the translation of ecological knowledge about fragmentation processes in the landscapes of Europe and its consequences for populations of natural species into policy relevant information.

Originally, the main goal of ecological networks was to conserve biodiversity by maintaining and strengthening the integrity of ecological and environmental processes; and to counter the above effects by linking fragmented ecosystems with each other in order to promote exchange between populations of species and to enable the migration and spread of species. As a conservation approach, ecological networks are characterized by two generic objectives, namely: 1) maintaining the functioning of ecosystems as a means of facilitating the conservation of species and habitats; and, like GI, 2) promoting the sustainable use of natural resources in order to reduce the impacts of human activities on biodiversity and/or to increase the biodiversity value of man-managed landscapes [12].

The concept of ecological networks is implicit in a variety of international conventions (e.g. Ramsar Convention, Bern Convention), European agreements (Habitats and Birds Directives) and related policy implementation (Natura 2000 and Emerald Networks). It has now become operational in a number of national and European strategies [13]. The development of a European Ecological Network formed one of the priorities and activities of European nature conservation under the Pan-European Biological and Landscape Diversity Strategy (PEBLDS) which was endorsed by 54 European countries in Sofia, in 1995.

Ecological network maps and strategies have also been established at country level and around trans-boundary sites and site complexes. The Natura 2000 site network is now well developed across the European Union member states and the Emerald network constitution process is making progress at a pan-European level, particularly in the Western Balkans, Central and Eastern Europe, South Caucasus, as well as Norway and Switzerland [14]. Together with other networks of protected sites that stem from international, national or regional arrangements, they provide the basis for planning and joint action, and a backbone of GI across the European Continent.

At regional and local level many planning authorities have applied the principles of ecological connectivity to spatial planning and strategies – and these are the levels where actual implementation is now taking place., The latter have often included a significant level of stakeholder and public involvement and participation in the planning process.

While the PEEN served as an umbrella policy initiative, the elaboration of national level ecological networks provided a route towards more concrete implementation. In recent years several such projects have been implemented – including, perhaps most recently, the Macedonian National Ecological network (MAKNEN) [15].

An important lesson learned from the process of planning and implementing ecological networks, which will also be crucial for the success or failure of the GI approach, is that policy on its own does not deliver action on the ground unless supported by funding. PEBLDS, for example, was able to provide both financial and political support to the progress of the PEEN while at national level such work was often carried out with the support of project funding (e.g. Croatia – LIFE III, Macedonia – BBI Matra Fund).

However, national ecological networks are unlikely to function effectively unless they cross national boundaries. Many ecosystems, for instance in mountain ranges (the Alps, the Pyrenees, the Carpathians) and along river basins (the Rhine, the Danube), extend beyond national boundaries and are part of the EU's shared natural and cultural heritage and identity. There are already several good examples of cross boundary and regional ecological networks targeting these ecosystems (e.g. within the framework of the Alpine convention, Carpathian Convention, in the Dinaric Arc, European Green Belt, etc.). The new GI Strategy therefore looks at the possibility of developing a trans-European GI (TEN-G) initiative, similar to that already in place for large-scale EU transport (TEN-T) and energy (TEN-E) networks [16].

3 Green Infrastructure – Future Perspective

Whilst the term "Green Infrastructure" has in the past been used to describe natural, connected habitat within urban areas, with the launch of the EU 2020 European biodiversity headline target and 2050 vision [9], it has been taken forward by the European Commission as a much broader and much more complex concept which is emerging as a new and potentially influential policy instrument. In its Strategy and Roadmap to resource efficiency [17] the Commission also included a strategy on GI for which it received a mandate from the European Council and the European Parliament, "as a contribution to further integrating biodiversity considerations into other EU policies" [18]. Finally the GI Strategy was adopted on 6 May 2013 in the form of the Communication on "Green Infrastructure (GI) – Enhancing Europe's Natural Capital".

The GI Strategy defines Green Infrastructure as: *a strategically planned network of natural and semi-natural areas with other environmental features designed and managed to deliver a wide range of ecosystem services. It incorporates green spaces (or blue if aquatic ecosystems are concerned) and other physical features in terrestrial (including coastal) and marine areas. On land, GI is present in rural and urban settings.*

The Strategy elaborates further that GI is a successfully tested tool for providing ecological, economic and social benefits through natural solutions. It helps us to understand the value of the benefits that nature provides to human society and to mobilise investments to sustain and enhance them. It also helps to avoid relying on infrastructure that is expensive to build when nature can often provide cheaper, more durable solutions. Many of these create local job opportunities. GI is based on the principle that protecting and enhancing nature and natural processes, and the many benefits human society gets from nature, are consciously integrated into spatial planning and territorial development. Compared to single-purpose, conventional infrastructure (also referred to as "grey" infrastructure), GI has many benefits. It is not a constraint on territorial development but promotes natural solutions if they are the best option. It can sometimes offer an alternative, or be complementary, to standard grey solutions [10].

It should be noted that the GI strategy is built on a number of different theoretical and conceptual starting-points in the fields of landscape ecology, conservation biology and nature protection. This results in the use of potentially inconsistent terminology which is perhaps inevitable when trying to combine different disciplines into a new single approach acceptable to practitioners from different backgrounds. However, it is generally accepted that GI includes the following elements [19]:

- Protected areas, such as Natura 2000 sites;
- Healthy ecosystems and area of high nature value outside protected areas such as floodplain areas, wetlands, coastal areas, natural forests etc...;
- Natural landscape features such small water courses, forest patches, hedgerows which can act as eco-corridors or stepping stones for wildlife;
- Restored habitat patches that have been created with specific species in mind e.g. to help expand the size of a protected area, increase

foraging areas, breeding or resting for these species and assist in their migration/dispersal;

- Artificial features such as eco-ducts or eco-bridges, that are designed to assist species movement across insurmountable landscape barriers;
- Multifunctional zones where land uses that help maintain or restore healthy biodiverse ecosystems are favoured over other incompatible activities;
- Areas where measures are implemented to improve the general ecological quality and permeability of the landscape;
- Urban elements such as green parks, green walls and green roofs, hosting biodiversity and allowing for ecosystems to function and deliver their services by connecting urban, peri-urban and rural areas;
- Features for climate change adaptation and mitigation, such as marshes, floodplain forests and bogs - for flood prevention, water storage and CO_2 intake, giving space to species to react to changed climate conditions.

As GI clearly still has some form of coherent ecological network at its core, it would seem prudent to take into account and build further on the work that has already been done at various geographical levels in order to define areas of existing and potential ecological connectivity. Below the level of ecological corridors that cross within and between countries, this includes the green and blue veining that makes up the patchwork quilt of traditionally managed multifunctional landscapes.

In many ways the lessons learned through the years of work on building ecological networks can and should be applied when thinking about GI as most of the issues remain the same. There is still a need to integrate the ecological network concept, which includes coherence and connectivity, into the development of GI, and equally into spatial and other infrastructure planning. Further, GI should also rely on existing policy instruments (such as the Natura 2000 and EU Water Framework Directive) which offer the potential for strengthening ecological networks to be exploited to the full in their implementation. Furthermore the EU Common Agricultural Policy presents an opportunity for new measures to be introduced that will benefit connectivity and contribute to making it a common practice to incorporate green infrastructure into the European agricultural landscapes. GI can make a significant contribution to the implementation of many of the EU's main policy objectives, especially in relation to regional and rural development, climate change, disaster risk management, agriculture and forestry [23, 24]. The new GI strategy advocates the full integration of a Green Infrastructure into these policies so that it becomes a standard component of territorial development across the EU.

The EU's main policies and their accompanying financial instruments will be vital for mobilising the potential of EU regions and cities to invest in GI. EU financed interventions can help to change the underlying paradigm from one where economy and environment are seen as trade-offs to one where the synergies and co-benefits are increasingly appreciated. They will also enable decision makers, stakeholders and civil society to achieve complex policy objectives such as regional and rural development, as well as water, resource efficiency and biodiversity goals whilst at the same time promoting new business opportunities for Small and Medium Enterprises (SMEs); for instance through the planning, implementation and monitoring of GI initiatives. [16].

The step that GI can take beyond what has already been achieved (with ecological networks) is to provide further context for informing the important decisions that need to be made in relation to the planning and management of the wider countryside outside of protected areas and other special sites. Thus, the further consideration of issues such as ecosystem services, climate change adaptation and ecological resilience are at the core of the GI approach [20].

An Expert Working Group on GI was set up by the European Commission to provide recommendations on what the EC work on GI should tackle from the view of stakeholders, Member States and scientists. During their work, the Working Group identified a number of benefits that GO can offer. According to these, GI should provide environmental, economic and social benefits, mainly by encouraging partnerships; and a crucial element in achieving this is the active involvement of relevant stake- and resource holders on the ground. It should continue to promote integrated spatial planning by identifying multi-functional zones and by incorporating habitat restoration measures and other connectivity elements into various land-use plans and policies. It should definitely be addressing the healthy-functioning of ecosystems, their protection and the provision and sustainable use of ecosystem goods and services, while increasing their resilience by addressing mitigation and adaptation to climate change. More specifically the ecosystem services it can provide include:

- It is an effective and cost-efficient tool for absorbing and sequestering atmospheric carbon dioxide (CO_2).
- It is contributing to the minimization of risks of natural disasters, by using ecosystem-based approaches for coastal protection through marshes/flood plain restoration rather than constructing dikes.

- Efficient use of GI can reduce energy usage through passive heating and cooling; filtering air and water pollutants; decreasing solar heat gain; providing wildlife habitat; reducing the public cost of storm water management infrastructure and provide flood control; providing food sources; and stabilising soil to prevent or reduce erosion.
- It may contribute to landscape aesthetics, preservation of archaeological and cultural heritage, provision of accessible open spaces, sustainable transportation and energy, opportunities for environmental education and strengthen community sense for nature and quality of life [21].

Ecosystem services, and their contribution and value to human wellbeing, are now an important component of the modern policy agenda. However, there is also evidence that the condition of most services has decreased in last 50 years [25]. By strengthening and maintaining the good functioning of ecosystems, GI can promote the multiple delivery of ecosystem services. This applies equally to existing (semi)natural ecosystems, such as wetlands or floodplains, as well as to 'new' ecosystems, such as green roofs and vertical farms, and especially in relation to land that has been degraded and which often occurs in urban areas [7].

Ultimately, GI aims to contribute to the development of a more sustainable economy by investing in ecosystem-based approaches delivering multiple benefits, in addition to technical solutions, and mitigating adverse effects of transport and energy infrastructure. In other words the: "...ultimate aim (of GI) is to provide the framework for the territorial development of a green and low carbon economy" [20]. In many ways much of this was indeed the desired and intended objective for the future development of the ecological networks concept; leading to the conclusion that GI is a natural evolution of ecological networks. 'EUROPE 2020 - A strategy for *smart, sustainable* and *inclusive* growth' puts forward these three mutually reinforcing priorities which should, respectively, be based on: developing an economy based on knowledge and innovation; promoting sustainable growth through greater resource efficiency, greener and more competitive economies; and fostering high-employment economies for improved social and territorial cohesion. Sustainable growth means building a resource efficient, sustainable and competitive economy, exploiting Europe's leadership in the race to develop new processes and technologies, including green technologies, accelerating the roll out of smart information and communication technology grids, exploiting EU-scale networks, and reinforcing the competitive advantages of the EU.

Linked to the Trans-European Networks (TENs) vision, key issues that can provide solid economic evidence include the scope to use GI thinking and measure GI contributions to resource efficiency across multiple objectives. Relevant here and to achieving important goals in such areas are two, out of seven, flagship initiatives set out in the Strategy; the "Innovation Union" which aims to:

- complete the European Research Area, to develop a strategic research agenda focused on challenges such as energy security, transport, climate change and resource efficiency, health and ageing, environmentally-friendly production methods and land management, and to enhance joint programming with Member States and regions;
- improve framework conditions for business to innovate;

and a "Resource efficient Europe" flagship initiative which, among other things, aims to:

- mobilise EU financial instruments (e.g. rural development, structural funds, R&D framework programme, TENs, European Investment Bank (EIB)) as part of a consistent funding strategy, that pulls together EU and national public and private funding;
- establish a vision of structural and technological changes required to move to a low carbon, resource efficient and climate resilient economy by 2050 which will allow the EU to achieve its emissions reduction and biodiversity targets [22].

3.1 Green Infrastructure from a Stakeholders' Point of View

Issues related to the implementation of GI were summarised nicely by the participants of the 'Greening European Regions Conference – biodiversity as a boost for regional and local economy'[1] which took place in Oisterwijk, The Netherlands, on 12 December 2012. The main challenges and barriers for applying the GI approach and incorporating it into regional development, as envisaged by the participants, were threefold: 1) policy – lack of coordination between different levels of governance and between different sectors; 2) available resources – financial and human; and 3) awareness and knowledge among policy makers, stakeholders and the general public.

---

[1]The Conference was organised by ECNC and UNEP, the Province of Limburg, (B), and the municipality of Oisterwijk, in cooperation with the Province of Noord-Brabant (NL), the EU Committee of the Regions and ENCORE (http://www.regionsandbiodiversity.eu/node/10)

When these were considered in more detail the main barriers identified were the following:

- **Lack of policy coordination in decentralisation**: while the general trend in policy making is decentralisation and giving more and more independence and responsibilities to the lower levels of governance (regional and local), there is often a lack of coordination between these different levels in setting the same or complementary priorities in order to make the implementation more effective.
- **Integration and political agreement between different policy levels**: there is often confusion about which level of governance is responsible for what, especially when it comes to planning – which is a key for the delivery of green infrastructure.
- **Political commitment at all levels**: there is often lack of firm political commitment towards the implementation of GI. Policy objectives often change with every new election and the consequent arrival of new politicians. This is the case at all levels: EU, national, regional and local.
- **Lack of human and economic resources**: GI based solutions are often long term projects which require ensuring long-term stable funding as well as qualified and knowledgeable teams of people; maintaining this in the long term can often represent a challenge.
- **Lack of wish for GI**: very often GI is a topic important to only a few people and therefore not a priority issue and not very high on the policy agenda; this is related to the low awareness of this issue amongst politicians and policymakers.
- **Lack of knowledge (arguments)**: evidence based arguments should be available and used to show the benefits of GI to both decision-makers and the general public.
- **Awareness is lacking**: there is a need for raising awareness on the possibilities and benefits of a GI based approach among the relevant stakeholders (i.e. different sectors) and general public.

When discussing possible solutions for these issues some of the suggestions were:

- **Involve stakeholders**: relevant stakeholders should be involved in the implementation process in order to change their mind with proper arguments – it is often possible to agree on the common goals.
- **EU Policies Coordination**: could be a tool for coordination between different policy levels.

- **Demonstrate the explicit economic value of GI**: such explicit examples (including definitive economic figures) should be used to assist in briefing and convincing politicians of the effectiveness of the approach.
- **Find innovative solutions**: GI based solutions should not be more expensive than 'grey' infrastructure solutions; this requires a certain level of innovation. Once identified, such solutions should be promoted.
- **Use more local/regional examples when communicating to stakeholders**: it is much easier for people to relate to the area they are familiar with.
- **Public – private commitment is a good approach towards implementation of GI**: a number of such examples already exist and these should be used to learn from and promote the approach further.

A number of these suggestions have in the meantime been picked up and are being followed up by the European Commission in their actions to facilitate the deployment of GI across the EU member states.

4 Conclusions

The contribution of ecological networks, and subsequently, GI to the provision of ecosystem services, transition to a more resource efficient economy and to mitigation and adaptation in relation to the effects of climate change are important areas for research and subsequent articulation into policy. Quantifying the economic benefits of ecological networks and GI and making them explicit through interdisciplinary research is also a clear necessity – looking into the social, economic and ecological mechanisms, as well as at the maintenance of biodiversity and the ecological services they provide.

Further work can be carried out in relation to the full translation of the protected area networks into functional ecological networks and making them integral building blocks of the GI both at the level of policy and practice. In addition information about how to create actual ecological networks at the delivery level, particularly where this has involved stakeholder and public participation needs to be researched and made widely available. Knowledge transfer is needed as well as new knowledge especially in relation to the impact of changing environmental and land use conditions on species and habitats in the wider countryside.

Leadership has already been mentioned in the context of who has responsibility for ecological networks at European, regional, national and local levels. Issues remain the same when we are talking about GI. Linked to this is the issue of communication; specifically to politicians and decision makers within key sectors (such as spatial planning, transport, industry, etc.) but more widely to researchers, conservation practitioners, businesses and the interested public [1].

As a new policy concept, GI sets ambitious goals to bridge the gap between the different sectors and to integrate benefits for biodiversity with those for socio-economic interests, to improve delivery of ecosystem services, climate change mitigation and, more than that, it aims to promote innovative solutions and the use of same land for multiple purposes. However, it is still somewhat unclear how all of this will be achieved and the European Commission, supported by scientists and environmental NGOs, is therefore currently placing a particular emphasis on providing guidance documents to demonstrate the benefits of GI for different sectors. A subsequent challenge will be to measure success in the short to medium term through the development of appropriate indicators.

References

[1] Jones-Walters, L. and Čivić, K., Draft Action Plan on the future strategic development of the Pan-European Ecological Network (PEEN) for the period 2012–2020. Council of Europe, T-PVS/PA (2012) 12, 2012.

[2] EEA, Landscape fragmentation in Europe. EEA Report No 2/2011, Joint EEA-FOEN report, 2011.

[3] Jongman R.H.G., Bouwma I.M., Griffioen A., Jones-Walters L. and Van Doorn A.M., The Pan European Ecological Network: PEEN, Landscape Ecology 26: 311–326, 2011.

[4] Council of Europe, 3rd meeting Report. T-PVS/PA (2011) 13. Convention on the conservation of European wildlife and natural habitats, Group of Experts on Protected Areas and Ecological Networks, Sept. 2011.

[5] Council Directive 92/43/EEC on the conservation of natural habitats and of wild fauna and flora, Official Journal L 206, 22/07/1992 P. 0007–0050, May 1992.

[6] Directive 2009/147/EEC of the European Parliament and of the Council on the conservation of wild birds (codified version of Directive 79/409/EEC), Official Journal L 20, 26/01/2010 P. 0007–0025, Nov. 2009.

[7] European Commission, The Multifunctionality of Green Infrastructure, In-depth report. Science for Environment Policy, 2012.

[8] Convention on Biological Diversity, Aichi Biodiversity Targets. TARGET 11 - Technical Rationale extended (provided in document COP/10/INF/12/Rev.1), 2010.

[9] European Commission, Our life insurance, our natural capital: an EU biodiversity strategy to 2020. Communication from the Commission to the European Parliament, the Council, the Economic and Social Committee and the Committee of the Regions. COM(2011) 244 final, 2011.

[10] European Commission, Green Infrastructure (GI) — Enhancing Europe's Natural Capital, COM/2013/0249 final, May 2013.

[11] Pan-European Biological and Landscape Diversity Strategy, Pan-European 2020 Strategy For Biodiversity. 13th Meeting of the Council for the Pan-European Biological and Landscape Diversity Strategy, STRA-CO (2011) 2, 2011.

[12] Bennett, G. and Wit, P., The Development and Application of Ecological Networks: a Review of Proposals, Plans and Programmes. Amsterdam: AIDEnvironment, 2001. In: Snethlage, M., L. Jones-Walters (Eds.), Interactions between policy concerning spatial planning policy and ecological networks in Europe (SPEN – Spatial Planning and Ecological Networks). ECNC, Tilburg, the Netherlands, 2008.

[13] Jongman, R.H.G., Külvik, M and Kristiansen. I., European ecological networks and greenways. Landscape and Urban Planning, 68:305–319, 2004.

[14] Council of Europe, Final data summary of the Joint EU/CoE Programme on the setting-up of the Emerald network. T-PVS/PA (2012) 4. Convention on the conservation of European wildlife and natural habitats, Group of Experts on Protected Areas and Ecological Networks, 2012.

[15] Brajanoska R., K. Čivić, K., Hristovski, S., Jones-Walters, L., Levkov, Z., Melovski, Lj., Melovski, D. and Velevski, M., Background document on Ecological Networks - Project: Development of the National Ecological Network in FYR Macedonia (MAK-NEN), MES, Skopje, Republic of Macedonia; ECNC, Tilburg, the Netherlands, 2009.

[16] European Commission, Building a Green Infrastructure for Europe, 2013.

[17] European Commission, Roadmap to a Resource Efficient Europe. COM(2011) 571 final, 2011.

[18] Council of the European Union, EU Biodiversity Strategy to 2020 - Council conclusions, 2011.

[19] European Commission, Green Infrastructure Fact Sheet, 2010.

[20] Jones-Walters, L. and Čivić, K., Ecological Networks and green infrastructure. In Ferdinandova, V. (ed.), EU Environmental Policies and Strategies in South-Eastern Europe - Training guidelines for involving CSOs from SEE in implementation of EU nature-related legislation IUCN, Belgrade, Serbia, 14–35, 2012.

[21] Green Infrastructure Working Group of the European Commission, Recommendations, Nov. 2011.

[22] European Commission, Communication from the Commission: EUROPE 2020 - A strategy for smart, sustainable and inclusive growth, COM(2010) 2020, March 2010.

[23] European Commission, Guidelines on Climate Change and Natura 2000, 2013.

[24] Naumann, S., Anzaldua, G., Berry, P., Burch, S., Davis, M., Frelih-Larsen, A., Gerdes, H. and Sanders, M., Assessment of the potential of ecosystem-based approaches to climate change adaptation and mitigation in Europe, Final report to the European Commission, DG Environment, Contract no. 070307/2010/580412/SER/B2, Ecologic institute and Environmental Change Institute, Oxford University Centre for the Environment, November 2011.

[25] Carpenter, S.R., Mooney, H.A., Capistrano, A.J. et al., Science for managing ecosystem services: beyond the Millenium Ecosystem Assessment, Proceedings of national Academy of Science USA 106:1305–1312, 2009.

Biographies

Kristijan Čivić is Senior Project Manager at ECNC in the Programme Area Green Infrastructure. His main responsibilities are development and implementation of project proposals in the field of ecological networks, Natura 2000 and Green Infrastructure. He is also a licensed workshop facilitator and has a significant experience in preparing and giving training sessions on stakeholder involvement and participatory action planning. He has worked in nature conservation policy and practice for more than 10 years. Prior to joining ECNC, Kristijan worked for five years at the State Institute for Nature Protection in Croatia, a government agency that delivers expertise and knowledge to the relevant ministry responsible for nature conservation. As such he has experience of project management, national and international policies and regulations regarding nature conservation. He participated in the process of establishing the National Ecological Network in Croatia. He worked on several international projects in South-East Europe related to ecological networks and stakeholder involvement.

Lawrence Jones-Walters is presently Head of Biodiversity and Policy at Alterra and has held a number of senior roles, most recently Deputy Executive Director ECNC-European Centre for Nature Conservation (2007–2013). He has worked in nature conservation policy and practice and sustainability for nearly 30 years and has extensive experience of national and international project and programme management (Europe, Asia and North America). His long experience in the science policy arena includes work with the Council of Europe (Group of Expert on Protected Areas and Ecological Networks), UNEP, the European Commission, the European Environment Agency and the European Topic Centre – Biodiversity. He is a regular contributor to international forums, conferences and journals; including contributing to the development of the international policy agenda, for example, in relation to ecological networks and green infrastructure. He has expertise in a range of areas, for example, he has been involved in the 'Streamlining European 2010 Biodiversity Indicators' (SEBI2010) process and its successor; and the European Topic Centre for Biodiversity (he currently chairs the Management Committee of the Topic Centre); and is co-author of the EEA report (2012) on European Protected Areas: past, present and future. He has also worked as an expert providing input to the environmental component of the Eurostat headline indicators for sustainable development. He has experience in sustainable development, ecosystem services and business and biodiversity and has worked in regional economic development towards the achievement of environmental integration into the sustainable development agenda (including work with business and industry, SMEs, social inclusion and health issues). He gained his PhD in conservation management, has a MSc in Applied Entomology, an MBA, and has an extensive publications list.

Machine Utilization Rates, Energy Requirements and Greenhouse Gas Emissions of Forest Road Construction and Maintenance in Romanian Mountain Forests

A. Enache and K. Stampfer

University of Natural Resources and Life Sciences, Vienna, Austria
Corresponding Authors: {adrian.enache; karl.stampfer}@boku.ac.at

Received 4 July 2014; Accepted 27 November 2014;
Publication 19 March 2015

Abstract

The FAO and EU forest strategies advocate the use of forest resources in ways which minimize the impact on the environment and climate. However, in forests with poor accessibility, the environmental footprint of forest operations is significant due to the long timber extraction distances. Thus, improving the environmental performance of forest operations requires a well-developed forest infrastructure, specifically the density and quality of roads. The aim of this paper was to assess the environmental footprint of forest roads in terms of embodied energy and greenhouse gas emissions due to construction and maintenance. In this respect, life cycle assessment approach was used to develop an input-output model for benchmarking two case study areas, considering real machine utilization rates, fuel consumption and labor requirements. The forest road life cycle was set to 30 years. Direct energy requirements derived from the fuel consumed by the machinery were considered. Construction and maintenance required energy inputs of 490.9 MJ m^{-1} and 580.4 MJ m^{-1}, respectively about 36.6 kg CO$_{2eq}$ m^{-1} and 43.1 kg CO$_{2eq}$ m^{-1} emission rates in the two case study areas, while occupying productive land with forest roads triggered a loss of 3.95 kg CO$_{2eq}$ m^{-1} y^{-1} and 4.40 kg CO$_{2eq}$ m^{-1} y^{-1} during the life cycle of the forest

Journal of Green Engineering, Vol. 4, 325–350.
doi: 10.13052/jge1904-4720.445

road. However, the CO_{2eq} loss due to road construction and maintenance is insignificant when compared to the CO_{2eq} stored in the growing stock of the opened forest area. Terrain characteristics showed a strong influence on the amount of fuel consumption, required energy input and GHG emissions, leading to higher environmental burden and higher road construction costs.

Keywords: Emissions, energy efficiency, forest, greenhouse gases, LCA, road construction, Romania.

1 Introduction

The EU 20-20-20 targets on climate change and energy sustainability envisage 20% reduction of greenhouse gas (GHG) emissions from 1990 levels and improving with 20% the energy efficiency by year 2020. Forests and their sustainable management play a major role in the reduction of GHG emissions level and in carbon storage in forest biomass (Kilpeläinen et al. 2011). The FAO and EU forest policy framework promote a holistic approach to the challenges of the entire forest value chain for adapting forests to climate change and for reducing the environmental footprint of forest operations within the framework of a low carbon economy. However, in forests with poor accessibility, the environmental footprint of forest operations is significant due to the long timber extraction distances.

Romanian forests cover 6.65 million ha (29% of the total land area; Abrudan et al. 2009) and have a poorly developed and unevenly distributed infrastructure (road density 6.5 m ha^{-1}; Olteanu 2008). Thus, skidding is the main method of timber extraction and the mean skid distance is about 1.8 km at national level (Popovici et al. 2003). Consequently, the environmental footprint of forest operations is high, while the productivity in timber harvesting and extraction is rather low (Borz et. al 2013; Enache et al. 2013). The average annual growth of Romanian forests is about 37 million m^3, the annual allowable cut (AAC) is 22.3 million m^3 and the average annual removal is about 17.0 million m^3 (World Bank 2012). About 65% of the forests are located in mountain ranges, 55% are state-owned forests and 45% non-state forests. The underdeveloped forest infrastructure makes sustainable forest management challenging, with significant pressure and environmental footprint on the accessible forests. However, the net forest growth in the last decades was positive, ranging between 15–17 million m^3 each year, triggering a consequent increase of carbon storage

(World Bank 2012). This means there is significant potential for increasing the sustainable wood mobilization, which requires a well-developed forest infrastructure.

Timber harvesting and road engineering have the most visible environmental impact in the forest sector. The life cycle assessment (LCA) is a suitable tool for approaching such challenges of the wood supply chain and for producing reliable indicators on the environmental performance of systems and processes in the forest sector (Heinimann 2012). Meister (1995) emphasized that the environmental balance of forest operations is based on mass flows and energy balance of inputs and outputs of a system. In addition, Richter (1995) stressed that defining the boundaries of a LCA system is difficult, highlighting that wood supports most of the negative burdens of the forest management activities, while other ecosystem services of the forest management with direct positive effects on people and the environment do not. The environmental performance of silviculture operations, timber harvesting and transport have been extensively addressed in the literature (Berg and Lindholm 2005; Johnson et al. 2005; Klvac and Skouppy 2009; Michelsen et al. 2008; Seppala et al. 1998; Klvac et al. 2012), while only few studies have included forest roads in the analyzed system boundaries (Berg and Karjalainen 2003; Bosner et al. 2012; Whittaker et al. 2011). American researchers focused more on the effects of forest roads on soil erosion, sedimentation and water quality (Coulter 2004; Mills, 2006; Loeffler et al. 2008), whilst European researchers focused on the embodied energy and GHG emissions of forest roads (Heinimann and Maeda-Inhaba 2003; Heinimann 2012; Whittaker et al. 2011). Since the environmental impact of roads relate to their construction, maintenance and use (Treloar et al. 2004), complete LCA of forest roads is difficult and time consuming, depending on the system boundaries and on the number of inputs in the process analysis. Hence, a hybrid process based and input-output based LCA approach is recommendable for estimating project specific environmental impacts of forest roads (Treloar et. al 2004; Sharrard 2007).

In this context, considering the current concerns on the environmental performance of forest management activities (Abrudan et al. 2009; Karjalainen et al. 2003; Michelsen et al. 2008; Olofsson et al., 2011), the aim of this paper was to quantify the embodied energy, the loss of productive land and the GHG emissions from forest roads construction and maintenance through a comparative assessment of two case study areas. In this respect, a hybrid LCA approach was used, referring to the functional unit of road.

2 Material and Methods

2.1 Pre-Set Standards

This study focused on the energy requirements and GHG emissions of forest roads due to construction and maintenance during their life cycle. In this respect, the following standards were established: real utilization rates of machinery and consumption rates of materials and labor; real transport distances for machinery and materials; AAC of the forest area assigned to the forest roads; the life cycle of the forest roads was set to 30 years. The CO_{2eq} emissions were determined for a complete cycle of the diesel combustion process based on a stoichiometric combustion model (Heinimann 2012), for a net calorific value of diesel engines of 42.76 MJ kg^{-1} (Stanescu 2012) and the diesel density of 0.835 kg m^{-3} (Berg and Karjalainen 2003). The loss of productive land due to road construction was quantified for an average annual growth of 6.0 m^{3}ha^{-1}. For timber transport, the truck and trailer system with loading capacity of 25 m^3 was considered.

2.2 Input-Output LCA Model and System Borders

The energy efficiency and the emissions of greenhouse gases are important elements in LCA which focuses on the global warming potential (GWP) of a system. A typical LCA consists of setting goals and objectives, inventory analysis, impact assessment and interpretation of results (Heinimann 2012), while an optimal hybrid LCA model for construction should include economics, on-site activities, equipment, transportation, water, energy and social equity related aspects (Sharrard 2007). The hybrid LCA is based on deriving an input-output (I-O) LCA model and then case-specific LCA data for the analyzed system which are substituted in the I-O model (Treloar et al. 2004).

Heinimann and Maeda-Inaba (2003) showed how the concepts of commodities and activities and the oriented graph theory can be used in investigating I-O flows in forest roads construction. Figure 1 shows the LCA model of forest roads developed in this study for investigating the input-output flows of the road construction and maintenance works. This model refers only to the life cycle inventory of the roads and allows identification of material, energy, labor and emission flows within the system. The model was applied for both road construction and maintenance works, referring to activities such as: preparatory works (i.e. transport of machinery and material to the site, road bed clearance); embankments execution, drainage system and pavement finishing; and maintenance works (i.e. pavement reshaping;

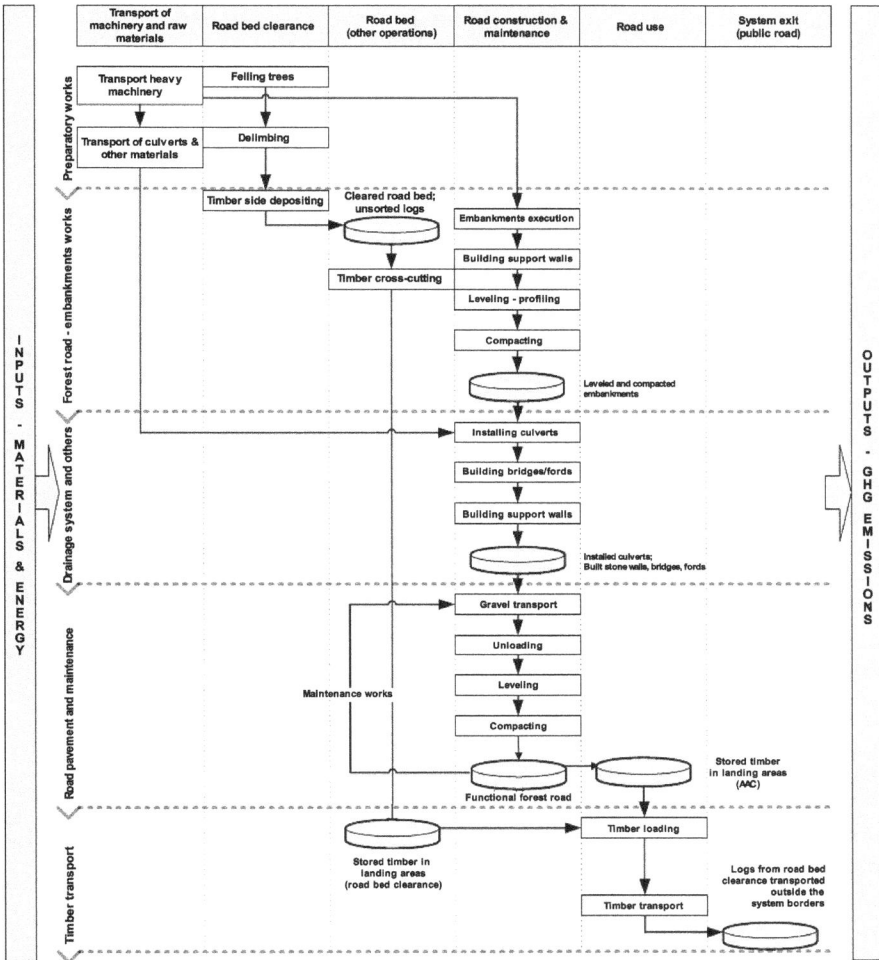

Figure 1 Life cycle I-O model for forest road construction and maintenance.

ditches reshaping). Direct energy requirements and greenhouse gas emissions were derived from the fuel consumed by machinery to carry out specific tasks (Whittaker et al. 2011), disregarding the energy and the emissions embodied in the machinery manufacture. The functional unit of the analyzed system set in this study was one meter of road.

Except for the timber cleared during the road construction, timber harvesting and transport were not included in analysis in this study. Accounting of the energy and emissions applied to timber harvesting and extraction with and without forest road will be approached in another study.

2.3 Building Technology Matrices

The quantification of the input and output flows for each phase of the LCA (Figure 1) were based on the technology matrices approach, using a system of linear equations which describe the flow of commodities into the system (Michelsen et al. 2008; Heinimann 2012), an example of which is presented in Table 1 for the phase of pavement works. The first row of the matrix shows the flow of labor necessary for a given process, taking into account the effective working time of a machine operator. The second row shows the fuel consumption rates of the machineries per productive system hour (PSH), which means the system includes both the machinery and the operator. The following rows were filled using the same reasoning. Thus, if a machine was not used in the system, all values in the row assigned to that machine were set to zero, except the diagonal value which was always set to value 1.

According to Heinimann (2012), assuming that each process can be scaled by a variable $x_{i(i=1 \div n)}$, the system of equations can be solved for the vector X $(x_1, x_2,..., x_n)$ if the total production of the system is known, that is vector Y, using the equations bellow.

Equation (1) $A \cdot X = Y$
Equation (2) $X = A^{-1} \cdot Y$

The economic performance of the systems was determined using a cost vector based on the machine hour costs computed with the FAO cost calculation scheme (Holzleitner, 2011), which was then multiplied with the performance vector X. Considering the direct correlation between the flow of commodities and their environmental footprint (Heinimann 2012), an environmental matrix similar to the technology matrix from the Table 1 was developed. This matrix was then multiplied with the performance vector X, and the environmental footprint vector of the analyzed system was thus determined. The technology matrix approach was used in benchmarking both forest road construction and maintenance works.

2.4 Case Study Areas (CSAs)

The research was conducted in *Lignum Forest Enterprise*, located in Bacau County (Romania), Eastern Carpathian Mountains (46°21′02″N, 26°20′42″E; Figure 2), in the surroundings of Accumulation Lake *"Valea Uzului"*

Table 1 Flow of commodities into the functional unit of road for pavement works

	Unit	X1	X2	X3	X4	X5	X6	X7	X8	X9	X10	X11	X12	X13	X14	
X1	Labor	hours	1	0	0	0	0	-1,23	-1,23	-1,23	-1,23	-1,23	-1,23	0	0	0	0
X2	Diesel fuel	liters	0	1	0	0	0	-12,9	-10,1	-10,0	-8,7	-8,65	-3,7	0	0	0	0
X3	Gasoline	liters	0	0	1	0	0	0	0	0	0	0	0	0	0	0	0
X4	Lubricants	liters	0	0	0	1	0	0	-0,24	-0,04	-0,07	-0,19	-0,05	0	0	0	0
X5	Chainsaw	PSH	0	0	0	0	1	0	0	0	0	0	0	0	0	0	0
X6	Excavator	PSH	0	0	0	0	0	1	0	0	0	0	0	0	0	-0,021	0
X7	Stone crusher	PSH	0	0	0	0	0	0	1	0	0	0	0	0	0	-0,066	0
X8	Grader	PSH	0	0	0	0	0	0	0	1	0	0	0	0	0	-0,032	0
X9	Front loader	PSH	0	0	0	0	0	0	0	0	1	0	0	0	0	-0,118	0
X10	Compactor	PSH	0	0	0	0	0	0	0	0	0	1	0	0	0	-0,025	0
X11	Dump truck	PSH	0	0	0	0	0	0	0	0	0	0	1	0	0	-0,204	0
X12	Trailer	PSH	0	0	0	0	0	0	0	0	0	0	0	1	0	0	0
X13	Timber lorry	PSH	0	0	0	0	0	0	0	0	0	0	0	0	1	0	0
X14	Preparing one unit of road	m	0	0	0	0	0	0	0	0	0	0	0	0	0	1	1
								Technology Matrix A									Y

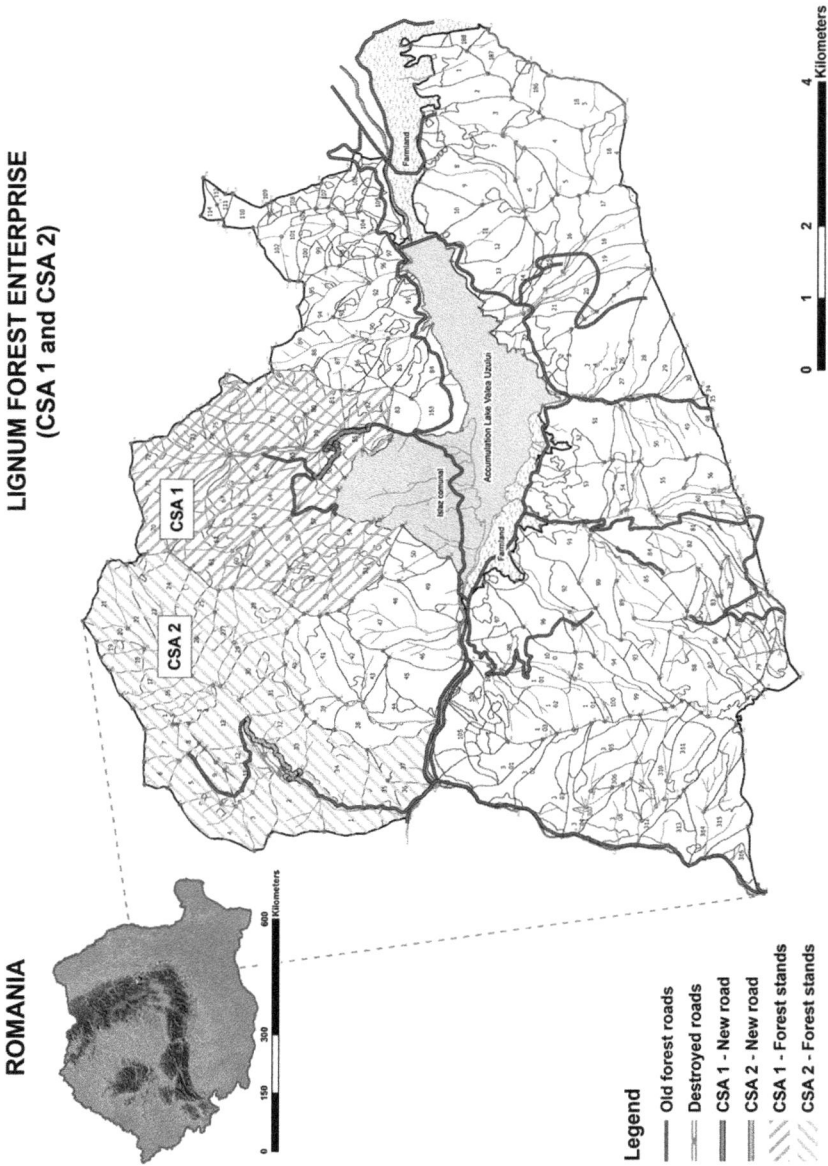

Figure 2   Location of the case study areas (CSA).

which provides drinking water for 27 communities with about 370 000 inhabitants.

The forest enterprise manages about 6 500 ha of mixed broadleaves-coniferous forests, of which 85% have mainly protective functions for water quality. The rotation of forest stands is about 100–110 years for conifers and 110–120 years for broadleaves. The bedrock is of Paleocene age, mainly sandstones, marl schist and alluvial formations, while the most common soil types are brown forest soils (75% of the area) and acid brown soils (25%). The density of forest roads is about 7.6 m ha^{-1}, with roads located mostly along the valleys. Timber extraction is done by tractors and skidders (65% of AAC), forwarders (21%), horse harnesses (8%) and cable yarders (6%). Two case study areas were selected for analysis: CSA 1 – *Forest Road Plopu-Lapos* and CSA 2 – *Forest Road Coporaia* (Table 2).

For each CSA, data of the following machineries used in road construction was collected from the records of the forest enterprise: chainsaw, excavator, stone crusher, grader, front loader, compactor, dump truck, trailer and timber lorry (Table 3). For road maintenance, data was gathered from the records of the maintenance works conducted in 2013 across the entire forest district for old valley forest roads, for the following machinery: backhoe loader, stone crusher, grader, compactor, dump truck and trailer.

2.4.1 CSA 1 – Forest Road "Plopu-Lapos"

The *Plopu-Lapos* forest road was built in 2013 and serves about 842 ha of forests (Figure 2), with an AAC of about 4 000 m^3 y^{-1}. The road has a length of 1.7 km, an average road bed width of 3.5 m and cross stations with an average width of about 7.0 m and 20 m length, located at intervals of 300 to 400 m. The permanent surface occupied by the road is about 1.02 ha, and the pavement structure is of 0.40 m thickness with gravel from on-site provenience. The road was built in moderate terrain conditions, using

Table 2 Key facts about case study areas

Item	CSA 1	CSA 2
Length of the new forest road (m)	1 707	1 968
Forest area served by the road (ha)	841.8	704.0
Current standing volume (m^3)	287 754	252 042
Estimated increment in 30 years (m^3)	151 376	142 032
Estimated gross standing volume after 30 years (m^3)	439 130	394 074
Estimated harvests in 30 years (m^3)	120 000	113 600
Estimated net standing volume in 30 years (m^3)	319 130	280 474

Table 3   Key facts of the machinery used in road construction and maintenance

Machinery	Producer/Model	Weight	Engine Power	Production year
Excavator	Hitachi Zaxis ZX225	22,5 to	110 kw	2006
Stone crusher	Hartl MT 503 BBV	31,8 to	186 kw	1999
Grader	O&K F156A	15,8 to	112 kw	2001
Compactor	Caterpillar CS583 C	14,7 to	72 kw	2000
Bulldozer	Liebherr LR 632	20 to	89 kw	1998
Front loader	Liebherr LR 632	22 to	132 kw	2004
Dump truck	MAN TG 3348	33 to/14 to	353 kw	2008
Timber lorry	Mercedes-Benz Actros 3348	33 to/16 to	350 kw	2007
Trailer	EMPL TLU 4X11	24 to	N/A	1998
Chainsaw	Husqvarna 372 XP	6,1 kg	3,9 kw	2012
Backhoe loader	Terex TX760B	6,8 to	69 kw	2005

a mixed cut-fill profile, with approximately 35% of the road length on low slope terrain and 65% of it on moderate slopes (Table 4). About 99% of the embankment works represented earth mass movements, while only 1% was rock mass movement, with no additional necessary works for stabilizing the slopes (Table 5). Before the new road was built no timber harvesting was

Table 4   Classification of forest roads by slope classes of the terrain in each CSA

Forest roads	Road CSA1		Road CSA2	
	Length (m)	Share (%)	Length (m)	Share (%)
Side slope classes of the terrain				
<25%	600,0	35%	500,0	25%
25–40%	1107,0	65%	510,0	26%
40–55%	0,0	0%	775,0	39%
>55%	0,0	0%	183,0	9%
Total road length (m)	1707,0	100%	1968,0	100%

Table 5   Characteristic of embankment works in each CSA

Forest roads	Road CSA1		Road CSA2	
	Volume (m^3)	Share (%)	Volume (m^3)	Share (%)
Total embankment works (m^3)	8179	100%	24097	100%
– earth mass movement	8104	99%	9880	41%
– rock mass movement	75	1%	14217	59%
Stabilizing support walls (m^3)	0	–	454	–

possible due to lack of access, currently the timber extraction is entirely done by skidders and forwarders.

2.4.2 CSA 2 – Forest Road "Coporaia"

The *Coporaia* forest road was built in 2011 and serves about 704 ha of forests (Figure 2), with an AAC of about 5 250 m^3 y^{-1}. The new road has a length of 2.0 km, an average road bed width of 3.5 m, cross station widths of about 7.0 m and 0.70 m thick gravel pavement from on-site provenience. The permanent surface occupied by this road is about 1.31 ha. The road was built in difficult terrain conditions; about 48% of the road length is located in steep and very steep terrain while 26% of the road length is on moderate slopes (Table 4). For steep and very steep slopes the road was built mostly in full bench profile, while for moderate slopes the mixed cut-fill profile was used. About 41% of the embankment works represented earth mass movements, while 59% was rock mass movement and about 450 m^3 of stones were necessary for stabilizing the slopes with supporting walls (Table 5). Before the road was built, timber extraction was done entirely with skidders on distances up to 3.5 km, while currently used extraction technologies are the skidders and forwarders (65% of the harvested volume) and the cable yarders (35%).

2.4.3 Road Maintenance

According to the Romanian regulations, the road maintenance works should be carried out regularly depending on the category of the forest road and of the amount of timber transported on it. Thus, considering the road network consists mainly of valley forest roads, *Lignum Forest Enterprise* performs road maintenance works at intervals of two years for each forest road. The maintenance works were split in two categories: one referring to pavement works (i.e. road bed and pavement structure reshaping, gravel replacement whenever necessary, leveling and compacting) and another one referring to the drainage system works (i.e. reshaping and cleaning the side ditches and the culverts). The collected data refers to maintenance works performed in 2013 on old valley forest roads which serve a total forest area of 750 ha with an AAC of 7 500 m^3 y^{-1} (35 % thinning, 50% final cuts and 15% sanitary cuts). The total length of repaired ditches was 7 000 m and the total length of reshaped road bed pavement was about 2 500 m. An additional amount of 85 m^3 of gravel was required for reshaping the pavement structure of the road. The gravel was transported from a local gravel deposit located 17 km away from the site.

3 Results

3.1 Machine Utilization Rates

3.1.1 Road Construction

The fuel consumption rates and the machine utilization rates for each phase of the road construction in CSA 1 are depicted in Table 6. For building one meter of road in CSA 1, 0.930 man-hours, 6.19 liters of diesel and 0.772 machine-hours were required. Out of the latter ones about 28% were excavator hours, 27% were dump truck hours and 15% were front loader hours.

The most intensive phases of road construction (in terms of labor, fuel consumption and machine utilization) were the embankments execution and the pavement works. The execution of embankments required about 26% of the labor, 39% of the fuel and 25% of the machinery utilization from the total amounts needed for building the road. For the pavement works, about 62% of the labor, 52% of the fuel and 60% of the machinery utilization were required.

The fuel consumption rates and the machine utilization rates for each phase of the road construction in CSA 2 are depicted in Table 7. For building one meter of road in CSA 2 were necessary about 8.59 liters of diesel and 1.084 machine-hours, out of which 31% were excavator hours, 23% were dump truck hours and 15% were front loader hours. Similar to CSA 1, the most intensive phases of the road construction in terms of labor requirements,

Table 6   Utilization rates of fuel, labor and machinery in CSA 1

Commodities		Preparatory Works	Embankment Works	Drainage system	Pavement works	Total Road Construction	
Labor	hours	0.086	0.239	0.032	0.573	0.930	hours
Diesel fuel	liter	0.192	2.420	0.336	3.242	6.191	liter
Gasoline	liter	0.058	0	0	0	0.058	liter
Lubricants	liter	0.031	0.035	0	0.039	0.105	liter
Chainsaw	PSH	0.042	0	0	0	0.042	PSH
Excavator	PSH	0	0.170	0.026	0	0.217	PSH
Stone crusher	PSH	0	0	0	0.066	0.066	PSH
Grader	PSH	0	0.016	0	0.032	0.047	PSH
Front loader	PSH	0	0	0	0.118	0.118	PSH
Compactor	PSH	0	0.008	0	0.025	0.033	PSH
Dump truck	PSH	0.004	0		0.204	0.208	PSH
Trailer	PSH	0.014	0	0	0	0.014	PSH
Timber lorry	PSH	0.026	0	0	0	0.026	PSH
Road unit	m	1	1	1	1	1	m

Table 7 Utilization rates of fuel, labor and machinery in CSA 2

Commodities		Preparatory Works	Embankment Works	Drainage system	Pavement works	Total Road Construction	
Labor	hours	0.187	0.462	0.024	0.411	1.084	hours
Diesel fuel	liter	0.375	5.080	0.313	2.818	8.586	liter
Gasoline	liter	0.170	0	0	0	0.170	liter
Lubricants	liter	0.090	0.058	0	0.042	0.190	liter
Chainsaw	PSH	0.123	0	0	0	0.123	PSH
Excavator	PSH	0	0.308	0.024	0	0.337	PSH
Stone crusher	PSH	0	0	0	0.075	0.075	PSH
Grader	PSH	0	0.022	0	0.004	0.025	PSH
Front loader	PSH	0	0	0	0.106	0.158	PSH
Compactor	PSH	0	0.014	0	0.048	0.062	PSH
Dump truck	PSH	0.006	0	0	0.173	0.245	PSH
Trailer	PSH	0.018	0	0	0	0.018	PSH
Timber lorry	PSH	0.040	0	0	0	0.040	PSH
Road unit	m	1	1	1	1	1	m

fuel consumption and machinery utilization in CSA 2 were the embankments execution and the pavement works. The execution of embankments required 43% of the labor, 59% of the fuel and 43% of the machine-hours from the total amounts needed for building the road, while the execution of pavement finishing required 38% of the labor, 33% of the fuel and 38% of the machinery utilization.

3.1.2 Road Maintenance

The utilization rates of the machinery used in one road maintenance operation are depicted in Table 8. About 0.5 liter of fuel and 0.072 machine-hours were required for maintaining one meter of road. Hence, during the entire life cycle of the forest road, maintenance works for one meter of road would require about 7.5 liters of fuel and 1.073 machine-hours utilization. The maintenance works of the drainage systems (i.e. reshaping the side ditches and cleaning the culverts) consumed about 60% of the total labor and fuel required for road maintenance.

3.2 Cost Appraisal

Table 9 shows the structure of the road construction and maintenance effort by type of costs. The total road construction costs were 88.2 € m^{-1} in CSA 1, respectively 119.6 € m^{-1} in CSA 2. The costs reported for road

Table 8   Machinery utilization rates for one process of road maintenance

Commodities	Pavement works	Drainage system	Total Road Maintenance	
Labor	0.029	0.043	0.072	hours
Diesel fuel	0.192	0.308	0.500	liter
Stone crusher	0.002	0.000	0.002	PSH
Grader	0.006	0.000	0.006	PSH
Backhoe loader	0.000	0.043	0.043	PSH
Compactor	0.003	0.000	0.003	PSH
Dump truck	0.016	0.000	0.016	PSH
Trailer	0.002	0.000	0.002	PSH
Road unit	1	1	1	m

Table 9   Structure of the road construction and maintenance costs

Cost types	Road construction CSA 1		Road construction CSA 2		Road maintenance	
	€/m	%	€/m	%	€/m	%
Machinery	22.0	25	37.7	32	1.5	51
Fuel	7.0	8	10.0	8	0.7	24
Labor	48.7	55	67.5	56	0.7	25
Materials	10.5	12	4.4	4	0	0
Total costs (€/m)	88.2	100	119.6	100	2.9	100

maintenance, respectively 2.91 € m^{-1}, are those required for performing one operation. Regarding the road construction, in both CSAs, the labor was the most intensive cost factor, representing about 55% (CSA 1) and 56% (CSA 2) of the total costs, respectively. The second most important cost factor in road construction was the utilization of machineries, with a share of 25% in CSA 1 and 32% in CSA 2 from the total costs. The most important cost factor in road maintenance was the machinery with about 51% of the total maintenance costs, while labor and fuel consumption had similar shares from the total costs, respectively 24% and 25%.

Figure 3 reveals the costs with preparatory works, drainage system execution and pavement works were similar in both CSAs, while the execution of embankments was significantly more costly in CSA 2 than in CSA 1, due to the steeper terrain and hence higher amounts of earth and rock excavations. In respect of road maintenance costs during the life cycle of the road, 60% of the costs are necessary for maintaining the pavement structure of the road and 40% of the costs for maintaining the drainage system. Considering one maintenance operation is carried out in average once at each two years, this means the yearly road maintenance costs are about 1.45 € m^{-1}. In addition, taking into account an yearly interest rate of 3.5%, the total maintenance costs

Road construction and maintenance costs (€ m⁻¹)

Road construction CSA 1   Road construction CSA 2   Road Maintenance

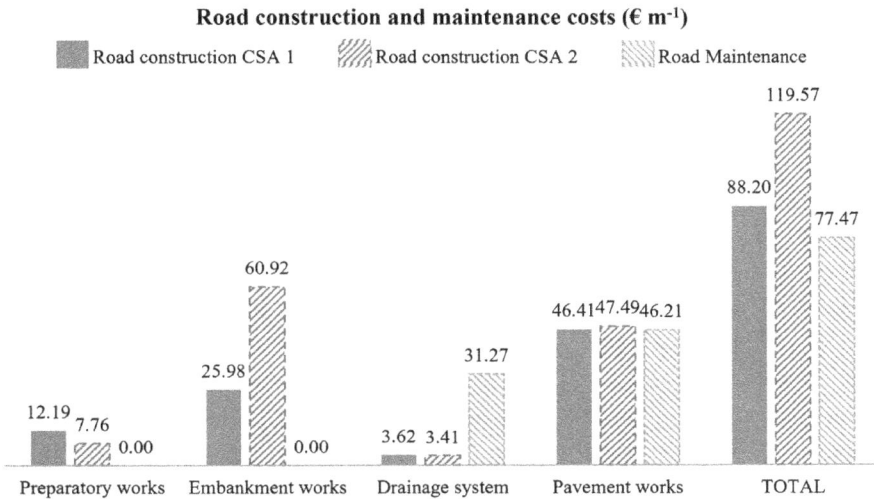

Figure 3   Forest road costs by category of works.

during the life cycle of the forest roads would be 77.5 € m^{-1} (Figure 3). This means that in moderate terrain conditions (i.e. CSA 1 - slopes below 40%) the initial investment costs in road construction would be almost equaled by the maintenance costs (about 88% of the construction costs) during the life cycle of the road, whereas in difficult terrain conditions (i.e. CSA 2 - slopes above 40% and stony material), the maintenance costs of the road would represent about 66% of the initial construction costs.

Table 10 presents the utilization rates, the fuel consumption rates and the system hour costs of the machinery (including fuel and labor costs). Slight variations of the machinery system hour costs were noticed between CSA 1 and CSA 2, respectively between road construction and road maintenance. This was probably because of the effective utilization time of the machineries and due to different operators running specific machineries. It has to be noted the forest enterprise used both local labor and Austrian labor for operating the machinery, the latter case being much more expensive, but however with more experience than local operators.

3.3 Embodied Energy, GHG Emissions and Loss of Productive Land

The most energy intensive phases in road construction are the embankment and the pavement works in both CSAs (Figure 4).

Table 10   Total utilization rates, fuel consumption and costs of machinery

Machinery	Road construction						Road maintenance		
	CSA 1			CSA 2					
	Hours	Fuel (l/h)	Costs (€/h)	Hours	Fuel (l/h)	Costs (€/h)	Hours	Fuel (l/h)	Costs (€/h)
Excavator	370	12.9	138.1	664	13.1	142.0	–	–	–
Stone crusher	112	10.1	134.1	112	10.1	137.8	6	9.7	134.8
Grader	81	9.7	96.5	50	10.0	100.5	14	10.1	95.4
Compactor	57	9.1	142.7	122	8.4	144.9	7	8.7	112.8
Front loader	202	8.7	90.8	311	8.7	101.6	–	–	–
Backhoe loader	–	–	–	–	–	–	299	7.2	27.5
Dump truck	349	3.7	84.1	471	3.7	108.7	40	4.6	30.7
Timber lorry	51	5.4	30.3	79	5.0	21.6	–	–	–
Trailer	24	3.5	32.4	47	7.3	57.9	5	7.3	35.1
Chainsaw	72	1.4	7.7	242	1.4	8.8	–	–	–
Total	1318	–	–	2133	–	–	371	–	–

Embodied energy in forest roads (MJ m⁻¹)

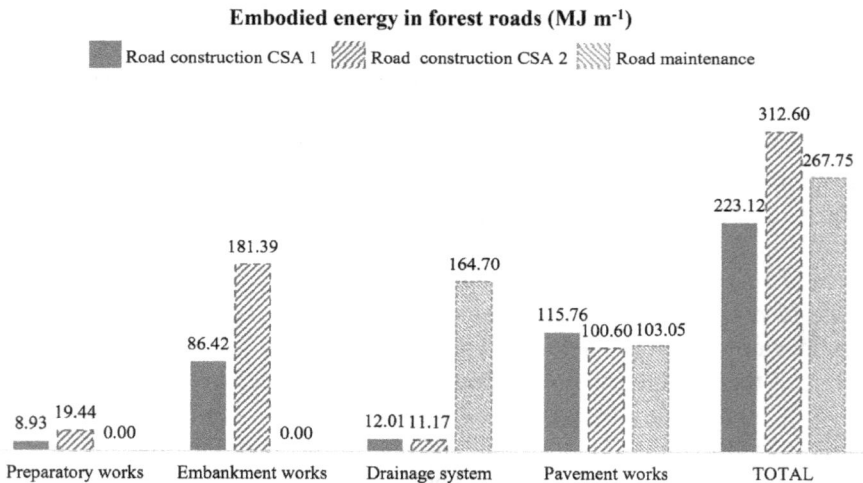

Figure 4   Energy requirements of road construction and maintenance.

Figure 4 reveals a significantly higher energy demand for the embankments execution in CSA 2 (181.4 MJ m^{-1}) compared to CSA 1 (86.4 MJ m^{-1}). This was due to the steeper terrain and more rock excavations in CSA 2 than in CSA 1 (Table 4 and Table 5), which required more machinery utilization. The total amount of energy required for road construction was 223.12 MJ m^{-1} in CSA 1 and 312.60 MJ m^{-1} in CSA 2 (Figure 4). The execution of pavement finishing was the most energy intensive phase in CSA 1, accounting for about 52% of the total energy input, while the most energy intensive phase in CSA 2 was the embankment execution, which accounted for about 58% of the total energy input.

One complete process of road maintenance required about 10.98 MJ m^{-1} for the construction of the drainage system and about 6.87 MJ m^{-1} for the pavement works. Although these figures might seem less energy intensive than the road construction, due to the repetition of this process at regular intervals during the entire life cycle of the road, the energy requirements of the maintenance works might equal or outweigh the energy input required in road construction: 164.70 MJ m^{-1} for drainage system and 103.05 MJ m^{-1} for pavement works. Therefore, the total energy embodied in forest roads due to construction and maintenance would beabout 490.87 MJ m^{-1} in CSA 1 and 580.35 MJ m^{-1} in CSA 2. Considering the allowable cut of each CSA and the life cycle of the forest roads, this means an energy input per cubic meter of timber harvested of about 7.0 MJ in CSA 1, respectively 7.3 MJ in CSA 2.

In what concerns the global warming potential of the forest road construction and maintenance, Table 11 shows the emission rates of $CO_{2eq.}$ per meter of road.

Forest road construction required about 16.6 kg CO_{2eq} m^{-1} in CSA 1 and about 23.0 kg CO_{2eq} m^{-1} in CSA 2. In both cases, the embankment and pavement works accounted together for more than 90% of the total GHG emissions. Considering the AAC and the life cycle of the roads, this means road construction has an environmental footprint of 0.236 kg CO_{2eq} in CSA 1, respectively of 0.251 kg CO_{2eq} in CSA 2 per cubic meter of timber harvested. From the emissions point of view, road maintenance has a lower environmental footprint per one process as such, requiring 0.515 kg CO_{2eq} m^{-1} for maintaining drainage systems and 0.823 kg CO_{2eq} m^{-1} for maintaining the pavement structure. However, due to the repeated interventions during the road life cycle, the CO_{2eq} emission rates of road maintenance are comparable to those

Table 11   Emission rates of CO_{2eq} from road construction and maintenance

GWP emissions of roads		Preparatory works	Embankment works	Drainage system	Pavement works	Total
$CO_{2eq.}$ (kg m^{-1})	CSA 1	0.514	6.479	0.900	8.679	16.573
	CSA 2	1.003	13.599	0.837	7.542	22.983
	Maintenance	0	0	12.345	7.725	20.070

of the initial road construction, one meter of maintained road requiring about 20.1 kg CO_{2eq}. In this study this means that in moderate slope conditions, the level of CO_{2eq} emissions from road maintenance works exceeds the emission levels from road construction in CSA 1 with about 21% and represent about 87% of the road construction requirements in difficult terrain conditions (CSA 2). Thus, road construction and maintenance works combined require about 36.6 kg CO_{2eq} m^{-1} in CSA 1 and 43.1 kg CO_{2eq} m^{-1} in CSA 2, respectively, which means emission rates of about 62.5 t CO_{2eq} in CSA 1 and 84.8 t CO_{2eq} in CSA 2 during the life cycle of the road. Considering the allowable cut in each case study area, these would mean about 0.521 kg CO_{2eq} in CSA 1 and 0.471 kg CO_{2eq} in CSA 2 per cubic meter of timber harvested during the road life cycle.

The permanent surface occupied by the road bed was 10 219 m^2 in CSA 1 and 13 108 m^2 in CSA 2 (Table 12). The loss of productive land due to road construction was about 5.07 m^2 y^{-1} per cubic meter of wood in CSA 1, respectively 5.82 m^2 y^{-1} in CSA 2. Considering the mean annual growth of forests in the study area (6 m^3 ha^{-1} y^{-1}) and that one cubic meter of wood binds about 1.1 tones CO_{2eq} from the atmosphere (Hasenauer, 2014), this means about 6.74 t CO_{2eq} in CSA 1 and 8.65 t CO_{2eq} in CSA 2 are not bound each year due to the loss of productive forest land. Reporting these figures to the road unit, it means that occupying productive forest land with forest roads requires 3.95 kg CO_{2eq} m^{-1} y^{-1} in CSA 1 and 4.40 kg CO_{2eq} m^{-1} y^{-1} in CSA 2.

The CO_{2eq} emissions due to loss of productive land can be only partially compensated by the CO_{2eq} stored in the timber harvested from the road bed clearance (Table 12). For clearing the road bed, about 407 m^3 were harvested in CSA 1 and 297 m^3 in CSA 2, which is equivalent to 447.7 CO_{2eq} and 326.7 t CO_{2eq}, respectively. Considering that approximately 40% of the harvested timber is used for wood products and 60% as energy wood by

Table 12   Impact of road bed clearance on CO_{2eq} emissions

Index	Item	CSA 1	CSA 2
1	Cleared road bed surface (ha)	1.02	1.31
2	CO_{2eq} emissions due to loss of productive land (t CO_{2eq})	202.2	259.5
3	CO_{2eq} from timber harvest road bed (t), of which:	447.7	326.7
4	– stored in wood products (t CO_{2eq})	179.1	130.7
5	– emissions to atmosphere (t CO_{2eq})	268.6	196.0
6	CO_{2eq} balance of the road bed clearance (t CO_{2eq}) [(6) = (4) – (5) – (2)]	−291.7	−324.8

Table 13 Impact of lost productive land on CO_{2eq} emissions during road life cycle

Index	Item	CSA 1	CSA 2
1	Total CO_{2eq} of current standing volume (t CO_{2eq})	316 530	282 750
2	Total CO_{2eq} of timber harvested in 30 years (t CO_{2eq}), of which:	132 000	124 960-
3	– stored in timber products (t CO_{2eq})	52 800	50000-
4	– emissions in atmosphere by energy wood (t CO_{2eq})	79 200	84960-
5	CO_{2eq} of net standing volume in 30 years (t CO_{2eq})	351 040	308 520-
6	CO_{2eq} balance of the road bed clearance (t CO_{2eq})	−291.7	−324.8
7	CO_{2eq} balance of road construction and maintenance (t CO_{2eq})	−62.4	−84.8
8	CO_{2eq} balance of the opened forest area (t CO_{2eq}) $[(8) = (7) + (6) + (5) + (3) − (4)]$	324 285	273 150

the forest enterprise, this means during the life cycle of the forest roads approximately 130.7 t CO_{2eq} in CSA 2 and 179.1 t CO_{2eq} in CSA 1 can be stored in wood products, the rest being released back in the atmosphere through the burning process. Table 12 reveals that occupying productive forest land with roads means a net loss of 291.7 t CO_{2eq} in CSA 1 and 324.8 t CO_{2eq} in CSA 2.

Notwithstanding, the net CO_{2eq} emissions due to loss of productive land are insignificant when compared to the amount of CO_{2eq} stored in the growing stock of the opened forest area. Table 13 shows the balance of CO_{2eq} due to the loss of productive land occupied by the roads during their entire life cycle for the forest area opened by the road construction in both CSAs. The CO_{2eq} balance was calculated as an algebraic sum of the CO_{2eq} gains (i.e. current standing volume, increment during the life cycle of the road, storage in wood products) and CO_{2eq} losses (i.e. emissions in atmosphere by combustion of energy wood). The CO_{2eq} emissions due to machinery utilization in timber harvesting were not included in this analysis.

4 Discussions and Conclusions

Forest road construction is an intensive process in what concerns machinery utilization, labor required, energy input and GHG emissions. The most energy intensive processes in road construction reported in this study were the embankment and the pavement works, accounting for about 90% of the total energy requirements in each CSA. The most intensive energy consumers and CO_2 emissions generators were the excavator, the dump truck and the front loader, accounting for about 70% of the total necessary machine utilization hours in each case study area.

On the other hand, road maintenance works are also energy intensive. Although one event of road maintenance is not so energy demanding (about 17.85 MJ m^{-1}) compared to road construction, the total energy required for maintenance works during the life cycle of the road (267.75 MJ m^{-1}) outweighs with about 20% the energy requirements for road construction in CSA 1 and represents about 85% of these in CSA 2. Basically, this means that maintaining valley forest roads over a life cycle of 30 years is almost as much energy intensive as the road construction, especially because of the number of repeated interventions. This is due to the well-known fact that valley forest roads are susceptible to more damages than slope roads, due to their vicinity to the water courses, particularly during spring and heavy precipitation season (i.e. snow melting and torrential flows). Therefore, it would be better from the point of view of energy input and GHG emissions to reduce the number of maintenance operations. This could be the case of road networks with more slope roads rather than with valley roads.

The total energy embodied in forest roads (construction and maintenance) was 490.87 MJ m^{-1} in CSA 1 and 580.35 MJ m^{-1} in CSA 2, respectively. In comparison, Heinimann (2012) estimated energy input rates for road construction and maintenance between 315 and 735 MJ m^{-1} road, depending on the side slope variation, while Whittaker et al. (2011) reported energy requirements of 403 MJ m^{-1} for road construction and 102 MJ m^{-1} for road maintenance, including the requirements of machine manufacture and maintenance. However, all these figures should be cautiously interpreted, looking at the characteristics of each study layout (i.e. topographical conditions, definition of the system borders).

One particularly important observation is that in this study the amount of fuel consumed in road construction (6.19 liters m^{-1} in CSA 1 and 8.58 liters m^{-1} in CSA 2) was almost similar to the amount of fuel needed forroad maintenance (7.5 liters m^{-1}) during the life cycle of the forest road. Thus, it can be underlined that the quality of the planning process and of the construction of forest roads plays a crucial role in the future running costs of a road network. Slope forest roads are easier and less costly to maintain than valley roads. Loeffler et al. (2009) estimated fuel consumption rates for road construction in full bench profile varying between 7.7 liters and 18.9 liters per meter of road depending on the side slopes (between 50% and 90%), while Whittaker et al. (2011) reported about 4.7 liters of fuel for building one meter of road, in a case study from Scotland. The terrain conditions and the characteristics of the forest road built in CSA 2 of this study were closer to those assumed by Loeffler et al. (2009) and so were the results regarding the fuel consumption.

The outcomes of this study have showed that road construction and maintenance operations are important sources of GHG emissions. The management practices (i.e. slope roads versus valley roads, share of timber used for timber products versus bio-energy) might have an influence on the environmental footprint of forest roads. However, the quantity of CO_{2eq} emissions from clearing the road bed, building and maintaining the road is insignificant in the equation of the CO_{2eq} emissions balance over the road life cycle. Hence, increasing the density of forest roads with about 2.0 m ha^{-1} in each case study area is worth while from the point of view of GHG emissions balance. Comparable findings were reported in the literature. Loeffler et al. (2009) estimated CO_2 emissions from road construction between 20.9 kg m^{-1} and 51.5 kg m^{-1} depending on the side slope variation, while Whittaker et al. (2011) showed about 37.8 kg CO_2 were required for building and maintaining one meter of road. Heinimann (2012) reported CO_2 output rates of road construction and maintenance between 19 kg m^{-1} to 47 kg m^{-1} depending on the terrain side slope conditions. Terrain characteristics have showed a strong influence on the amount of fuel consumption, the required energy inputs and the GHG emissions in this study, too. It was showed that steeper slopes and stonier terrain finally lead to higher environmental burden (i.e. 43.1 kg CO_{2eq} m^{-1} in CSA 2 compared to 36.6 kg CO_{2eq} m^{-1} in CSA 1) and higher road construction costs (i.e. 120 € m^{-1} in CSA 2 compared to 88 € m^{-1} in CSA 1). However, road construction costs are still very high when compared to similar terrain conditions from other countries. For example, in Austria, the road construction costs may vary from 14 € m^{-1} and 100 € m^{-1} depending on the terrain slope and stoniness, while the average cost is about 35 € m^{-1} (Ghaffariyan et al. 2010).

The input-output LCA approach proved to be a useful tool for assessing the energy requirements and GHG emission levels of forest roads. Though, setting the system boundaries and the time scale, gathering and analyzing data represent challenging and time consuming tasks. A natural further step of this study would be the accounting of the energy and emissions of different harvesting systems in mountain regions with and without forest roads, in order to see the impact of forest infrastructure development on the environmental footprint of harvesting operations.

5 Acknowledgments

The authors would like to thank Hendrik Schubert and Constantin Vasilică from *Lignum Forest Enterprise* for their support in data collection and feedback during data analysis. We would also like to thank the two anonymous reviewers for their valuable comments and suggestions.

References

[1] Abrudan, I. V., Marinescu, V., Ionescu, O., Ioras, F., Horodnic, S. A., and Sestras, R. 2009. Developments in the Romanian forestry and its linkages with other sectors. *Notulae Botanicae Horti Agrobotanici Cluj-Napoca* 37(2):14–21.

[2] Borz, S. A., Dinulica, F., Birda, M., Ignea, G., Ciobanu, V. D., Popa, B. 2013. Time consumption and productivity of skidding silver fir (*Abies alba* Mill.) round wood in reduced accessibility conditions: a case study in windthrow salvage logging from Romanian Carpathians. *Annals of Forest Research* 56(2):363–375.

[3] Bosner, A., Poršinsky, T., and Stankić, I. 2012. Forestry and life cycle assessment. P. 139–160 in *Global perspectives on sustainable forest management*, Dr. Clement A. Okia (Ed.). InTech, ISBN: 978-953-51-0569-5.

[4] Berg, S. and Karjalainen, T. 2003. Comparison of greenhouse gas emissions from forest operations in Finland and Sweden. *Forestry* 76(3):271–284.

[5] Berg, S. and Lindholm, E.-L. 2005. Energy use and environmental impacts of forest operations in Sweden. *Journal of Cleaner Production* 13:33–42.

[6] Coulter, E. D. 2004. *Setting forest road maintenance and upgrade priorities based on environmental effects and expert judgment.* PhD Thesis, Oregon State University, Corvallis, 199 p.

[7] Enache, A., Kühmaier, M., Stampfer, K., Ciobanu, V. D. 2013. An integrative decision support tool for assessing forest roads options in a mountainous region in Romania. *Croatian Journal of Forest Engineering* 34 (1):43–60.

[8] Ghaffariyan, M. R., Stampfer, K., Sessions, J., Durston, T., Kuehmaier, M., Kanzian, C. (2010): Road network optimization heuristic and linear programming. *Journal of Forest Science* 56(3): 137–145.

[9] Heinimann, H. R. and Maeda-Inaba, S. 2003. Quantification of environmental performance indicators EPIS for forest roads. In: Proceedings of Austro 2003: *High Tech Forest Operations for Mountainous Terrain*, Schlaegl, Austria, 13 p.

[10] Heinimann, H. R. 2012. Life cycle assessment (LCA) in Forestry – State and Perspectives. *Croatian Journal of Forest Engineering* 33(2): 357–372.

[11] Holzleitner, F., Kanzian, C., and Stampfer, K. 2011. Analyzing time and fuel consumption in road transport of round wood with an onboard fleet manager. *European Journal of Forest Research* 130:293–301.

[12] Johnson, L. R., Lippke, B., Marshall, J. D., and Comnick, J. 2005. Life-cycle impacts of forest resource activities in the Pacific Northwest and Southeast United States. *Wood and Fiber Sciences* 37:30–46.

[13] Karjalainen, T., Pussinen, A., Liski, J., Nabuurs, G.-J., Eggers, T., Lapvetelainen, T., and Kaipainen, T. 2003. Scenario analysis of the impacts of forest management and climate change on the European forest sector carbon budget. Forest Policy and Economics 5: 141–155.

[14] Klvac, R., Fischer, R., and Skoupy, A. 2012. Energy use of and emissions from the operation phase of a medium distance cableway system. *Croatian Journal of Forest Engineering* 33(1):79–88.

[15] Loeffler, D., Jones, G., Vonessen, N., Healey, S., and Chung, W. 2008. Estimating diesel fuel consumption and carbon dioxide emissions from forest road construction. In: Proc. *Forest Inventory and Analysis (FIA) Symposium 2008* RMRS-P-56CD, McWilliams et al. (Ed.), Fort Collins, 11 p.

[16] Meister, G. (1995): Eco-balances in the pulp industry: assessment of environmental impacts. P. 107–114 In: Proc. of the *International Workshop "Life cycle analysis – a challenge for forestry and forest industry"*, Hamburg, Germany.

[17] Michelsen, O., Solli, C., and Stromman, A. H. 2008. Environmental impact and added value in forestry operations in Norway. *Journal of Industrial Ecology* 12 (1): 69–81.

[18] Mills, K. (2006): Environmental performance measures for forest roads. P. 291–300 In: Proc. of the 29th *Council on Forest Engineering Conference: "Working globally – sharing forest engineering challenges and technologies around the world"*, Coeur d'Alene, Idaho.

[19] Olofsson, P., Kuemmerle, T., Griffiths, P., Knorn, J., Baccini, A., Gancz, V., Blujdea, V., Houghton, R. A., Abrudan, I. V., and Woodcock, C. E. 2011. Carbon implications of forest restitution in post-socialist Romania. *Environmental Research Letters* 6: 045202, 10 p.

[20] Olteanu, N. 2009. Considerations regarding the optimum density in Romanian forests. P. 701–708 In: Proc. of the *Biennial International Symposium "Forest and Sustainable Development 2008"*, Transilvania University of Brasov.

[21] Popovici, V., Bereziuc, R., and Clinciu, I. 2006. Considerations regarding the enhancement of the forest road network for opening up the forests. *Revista Padurilor* 121(6):19–21.

[22] Richter, K. (1995): Life cycle analysis of wood products. P. 69–77 In: Proc. of the *International Workshop "Life cycle analysis – a challenge for forestry and forest industry"*, Hamburg, Germany.

[23] Seppala, J., Melanen, M., Jouttijarvi, T., Kauppi, L., and Leikola, N. 1998. Forest industry and the environment: a life cycle assessment study from Finland. *Resources, Conservation and Recycling* 23: 87–105.

[24] Sharrard, A. L. 2007. Greening Construction Processes Using an Input-Output-Based Hybrid Life Cycle Assessment Model. *PhD thesis,* Carnegie Mellon University Pittsburgh, Pennsylvania, 343 p.

[25] Stanescu, R. C. 2012. Research concerning production technologies and properties of biofuels for vehicles. *PhD Dissertation*, Transilvania University of Brasov, Romania.

[26] Treloar, G. J., Love, P. E. D., and Crawford, R. H. 2004. Hybrid life-cycle inventory for road construction and use. *Jour. of Constr. Eng. and Manag.* 130(1):43–49.

[27] Whittaker, C., Mortimer, N., Murphy, R., and Matthews, R. 2011. Energy and greenhouse gas balance of the use of forest residues for bioenergy production in UK. *Biomass and Bioenergy* 35: 4581–4594.

[28] World Bank 2012. Functional Analysis of the Sector Environment and Forests in Romania – Vol. 1 & 2. – Final Report, 66 p.

Biographies

Adrian Enache is a research associate at the Institute of Forest Engineering from the University of Natural Resources and Life Sciences (BOKU) –Vienna, Austria. His work focuses on efficiency gaps analysis and multiple criteria decision making in timber harvesting and road network planning in mountain forests (ARANGE project). He holds a doctor degree in forestry (*Transilvania* University of Brasov – UNITBV, Romania, 2013), a master degree in mountain forestry (BOKU, 2009), a certification in project management (CODECS, 2007) and a diploma engineer degree in forestry (UNITBV, 2005). Since 2011 he is enrolled in a PhD program at BOKU.

Karl Stampfer is professor at the Institute of Forest Engineering at BOKU, Vienna. His expertize is related to the multi-dimensional aspects of the wood supply chain (i.e. timber harvesting, road network planning, logistics, cable yarding, ergonomics). He holds a doctoral degree in forestry since 1996 (BOKU) and his habilitation was in 2002 (BOKU). He received several international awards and he is member in numerous international professional societies (i.e. IUFRO, Austrian Association for Agricultural Research, Croatian Academy of Forestry Sciences). Since 2003 he is the chairman of FORMEC - International Symposium on Forestry Mechanization.

Author Index

Keywords Index